FOR OUR CHILDREN

VIBS

Volume 215

Robert Ginsberg
Founding Editor

Leonidas Donskis
Executive Editor

Associate Editors

G. John M. Abbarno
George Allan
Gerhold K. Becker
Raymond Angelo Belliotti
Kenneth A. Bryson
C. Stephen Byrum
Robert A. Delfino
Rem B. Edwards
Malcolm D. Evans
Daniel B. Gallagher
Roland Faber
Andrew Fitz-Gibbon
Francesc Forn i Argimon
William Gay
Dane R. Gordon
J. Everet Green
Heta Aleksandra Gylling
Matti Häyry

Brian G. Henning
Steven V. Hicks
Richard T. Hull
Michael Krausz
Olli Loukola
Mark Letteri
Vincent L. Luizzi
Adrianne McEvoy
Peter A. Redpath
Arleen L. F. Salles
John R. Shook
Eddy Souffrant
Tuija Takala
Emil Višňovský
Anne Waters
James R. Watson
John R. Welch
Thomas Woods

a volume in
Values in Bioethics
ViB
Matti Häyry and Tuija Takala, Editors

FOR OUR CHILDREN
The Ethics of Animal Experimentation in the Age of Genetic Engineering

Anders Nordgren

Amsterdam - New York, NY 2010

Cover Design: Studio Pollmann

The paper on which this book is printed meets the requirements of "ISO 9706:1994, Information and documentation - Paper for documents - Requirements for permanence".

ISBN: 978-90-420-2804-3
E-Book ISBN: 978-90-420-2805-0
© Editions Rodopi B.V., Amsterdam - New York, NY 2010
Printed in the Netherlands

To Matilda, Johanna, and Camilla

CONTENTS

Preface		ix
ONE	Introduction: Animal Experimentation, Public Opinion, and Philosophical Debate	1
	1. Animal Experimentation	2
	2. Genetically Modified Animals	4
	3. Public Attitudes toward Animal Experimentation	7
	4. Public Attitudes toward Genetically Modified Animals	9
	5. The Philosophical Debate	10
	6. Outline	11
TWO	Five Ethical Prototypes of Animal Experimentation	13
	1. Human Dominion	14
	2. Equal Consideration of Interests	21
	3. Animal Rights	29
	4. Strong Human Priority	35
	5. Weak Human Priority	39
	6. A Spectrum of Views	45
THREE	The Case for "Weak Human Priority"	47
	1. Legal Regulation and the Five Prototypes	47
	2. Ethical Theory	49
	3. Key Metaphors	52
	4. Intrinsic *and* Relational Properties	54
	5. Reason *and* Feelings	56
	6. Impartiality *and* Special Obligations	59
	7. From "Is" to "Ought"	63
	8. Strengths and Weaknesses of the Five Prototypes	68
	9. Proposal: Weak Human Priority	69
	10. Differences Compared to Midgley's Version	76
	11. Moral Imagination and Imaginative Casuistry	78
	12. Moral Imagination in Animal Experimentation	82
FOUR	The Scientific Value of Animal Experimentation	85
	1. Animal Experimentation in Present-Day Basic and Applied Research	85
	2. Prototypical Cases of Scientifically Valuable Animal Experiments	90
	3. The *Con* Argument from Causal Disanalogy	92

	4. Objections to the *Con* Argument from Causal Disanalogy	98
	5. The 3Rs: Replacement, Reduction, Refinement	103
	6. Implications of the Five Prototypes for the 3Rs	104
	7. The 3Rs: Practical Implications	105
FIVE	Animal Welfare and Ethical Balancing	111
	1. Three Animal Welfare Concerns	111
	2. A Comprehensive Approach	117
	3. Conceptual Implications	118
	4. Ethical Implications	120
	5. Animal Sentience	127
	6. Animal Welfare in Animal Experimentation	134
	7. Ethical Balancing in Animal Experimentation	137
SIX	Genetically Modified Animals in Research	147
	1. Implications of the Five Prototypes	147
	2. Scientific Concerns	148
	3. Intrinsic Ethical Concerns	156
	4. Animal Welfare Concerns	165
	5. Ethical Trade-Off: Four Cases	174
	6. Conclusion	179
Works Cited		181
Index		195

PREFACE

This book is the combined result of two research projects. The first was "Transgenic animals in research: ethical aspects," funded by the Swedish Foundation for Strategic Research (the ELSA (Ethical, Legal, and Social Aspects) Program) and The Knut and Alice Wallenberg Foundation (Program of ethics research in connection to Swegene and WCN (Wallenberg Consortium North)). The focus of this project was the production and use of genetically modified animals in biomedical research. As a background to this study, I investigated the ethics of animal experimentation in general. In an empirical part of the project, my colleague Helena Röcklinsberg and I analyzed applications regarding genetically modified animals submitted to ethics committees on animal experimentation in Sweden. The second project was "On health and welfare in the worlds of animals and humans" led by Professor Lennart Nordenfelt and funded by the Swedish Council for Working Life and Social Research. It was a comparative study of human and veterinary medicine regarding the concepts of health and welfare. Within this project I had the opportunity to study the ethical implications of different conceptions of animal welfare.

In writing this book, I received valuable input from several scholars and scientists. I am especially grateful to Bo Algers, Ted Ebendal, Stefan Gunnarsson, Mats G. Hansson, Robert Heeger, Henrik Lerner, Lennart Nordenfelt, Kerstin Olsson, Helena Röcklinsberg, Peter Sandøe, and Jan Vorstenbosch. Note that the views presented are my own and not necessarily shared by the persons mentioned.

Anders Nordgren
Linköping University

One

INTRODUCTION: ANIMAL EXPERIMENTATION, PUBLIC OPINION, AND PHILOSOPHICAL DEBATE

Would you accept the suffering of a large number of mice in a series of animal experiments that aim to develop a cure for cancer that might save the lives of our children? Would you accept minor animal suffering only, or also moderate or even severe suffering? Would you accept animal experiments in order to find cures for diseases that are not life-threatening? Would you accept mice suffering in experiments in order to obtain basic biological knowledge? Would you accept the use of chimpanzees in animal experiments? Would you accept the genetic modification of animals so that they can be used as disease models?

Questions like these force themselves upon us. How should we respond as private citizens and as a community? Some would argue that the likelihood of finding a cure is extremely small. Some would say that it is imperative to conduct research in order to find cures for cancer, but that we should do this with alternative methods; we should never use animals. Defenders of animal experimentation, on the other hand, would maintain that animal experiments are necessary, and point out that almost all pharmaceuticals and therapies in modern medicine have been developed on the basis of animal experiments.

Let us have a brief look at the statistics. Within the European Union, for example, 10.7 million animals were used in experiments in 2002 (Commission of the European Communities, 2005, p. 4; note that France submitted statistics from 2001). We have good reasons to believe that a substantial portion of these animals were genetically modified, that is, their genomes had been modified by technical means. No corresponding numbers for the European Union are available, but in the United Kingdom more than a quarter of the animals used in 2003 were genetically modified animals (Nuffield Council on Bioethics, 2005, p. 293).

These facts raise several ethical and policy questions. Is this extensive use of animals at all justified? Are animals the last slaves in our society, as animal liberationists suggest? How painful are present-day animal experiments? What does a realistic view of their benefit imply? On balance, are the costs in terms of animal suffering too high? Are there realistic alternatives to animal experiments? Does the use of genetically modified animals give rise to any special ethical problems? What are the advantages of using genetically modified animals in research?

These questions indicate that animal experimentation is among the most controversial issues raised by modern science (see, for example, LaFollette and Shanks, 1996; Paul and Paul, 2001; Armstrong and Botzler, 2003; Nuffield Council on Bioethics, 2005). But if animal experiments are controversial generally speaking, the production and experimental use of genetically modified animals are even more so (see, for example, Reiss and Straughan, 1996; Sherlock and Morrey, 2002; Nuffield Council on Bioethics, 2005). To some people, genetic "manipulation" is a "violation of natural order" or even a matter of "playing God."

In this book, questions like these will be discussed in detail. The overall objective is to develop a social ethic—that is, an ethical view that society should accept—of animal experimentation in biomedicine in general and research involving genetically modified animals in particular. So, a societal perspective instead of a personal one will be predominant. The special focus on genetically modified animals in research might need some explanation. I have two reasons for this focus. The first is that although my discussion will commonly be carried out in quite general terms, I need some concrete examples. Research involving genetically modified animals provides such examples. The second reason is that the issue of genetically modified animals is especially "hot" at the present time and deserves special attention. Ours is "the age of genetic engineering."

1. Animal Experimentation

What is an animal experiment? Different definitions have been suggested. Within the European Union, an "animal experiment" is defined as

> any use of an animal for experimental or other scientific purposes which may cause it pain, suffering, distress or lasting harm, including any course of action intended, or liable, to result in the birth of an animal in any such condition, but excluding the least painful methods accepted in modern practice (i.e. 'humane' methods) of killing or marking an animal; an experiment starts and ends when no further observations are to be made for that experiment; the elimination of pain, suffering, distress, or lasting harm by the successful use of anaesthesia or analgesia or other methods does not place the use of an animal outside the scope of this definition (Council Directive 86/609/EEC).

We see here that it is not sufficient that an experiment is carried out for a scientific purpose. It is also necessary that the experiment may cause animal pain, suffering, distress or lasting harm. To simplify, an animal experiment is not an animal experiment in the European Union sense if it does not imply that the animals suffer. However, the experiment is an animal experiment in this technical sense even if the animals do not actually suffer because some measures are taken to prevent, reduce, or eliminate the suffering by, for example, anesthesia or analgesia.

Other definitions can also be found. In the animal welfare legislation of my own country—Sweden—where the presence of a scientific purpose is sufficient for an experiment involving animals to be called an animal experiment. Animal suffering is not necessary (Animal Welfare Act, 1988 (with later revisions), Section 19).

In this book, I will use the broader definition including also experiments that do not involve any animal pain or suffering. In a discussion of ethics, we have no reason to adopt a narrow approach. For instance, some people may argue that animal experimentation may imply a violation of animal integrity or a hindrance of animals to lead a natural life even if no pain is inflicted on these animals.

Of importance is also how the term "animal" is defined. For example, the European Union directive defines it as

> any live non-human vertebrate, including free-living larval and/or reproducing larval forms, but excluding foetal or embryonic forms (Council Directive 86/609/EEC).

Definitions that include all non-human vertebrates hold generally in legal regulation all over the world, although we find at least one important exception, namely the Animal Welfare Act of the United States (1985). This act does not include mice, rats, and domestic birds among "animals." In the special guidelines for federally funded research, however, these species are regulated (Public Health Services, 1986).

Even some species of invertebrates may be sentient, that is, have some consciousness and ability to feel pain. Many people view *Octopus vulgaris* and cephalopods as sentient species. It could therefore be argued that they should count as "animals" in the regulation of animal experimentation. In some countries, this is also the case. In the United Kingdom, for example, *Octopus vulgaris* is included (Animals (Scientific Procedures) Act, 1986). In this book, I will include also sentient invertebrates.

In the European Union definition unborn animals are not viewed as "animals" in the technical sense. However, in some countries—for example, the United Kingdom—fetal, larval and embryonic forms that have reached specified stages of development are included (Animals (Scientific Procedures) Act, 1986). In the ethical discussion of this book, I will talk about post-natal sentient animals if not stated otherwise.

In the European Union, the most commonly used species in animal experimentation is the mouse (51%) followed by the rat (22%). Primates represent less than 0.1% (Commission of the European Communities, 2005, p. 5; these numbers concern 2002, except for France, which submitted statistics from 2001). With this in mind, the main focus in this book will be on the use of mice and rats.

Scientists commonly put forward three main reasons for conducting animal experimentation. One reason is to improve human health by developing diagnostics and treatments for different diseases. By conducting animal experiments better knowledge of disease causes and disease processes are obtained, leading to better diagnostic techniques and pharmacological, surgical or other therapies. In pharmaceutical research, animal experiments are used to test how a new drug works and how effective it is. They are also used to test the toxicity of new or existing drugs and of other chemicals.

Another reason is to obtain basic knowledge in biology, medicine, animal research, psychology, and so on. We know from the history of medicine that animal experiments in basic research have sometimes led to benefits for future patients even in cases in which no such medical application was foreseen.

A third main reason is that of improved animal health. Many experiments in veterinary medicine have this purpose. A key concern is improved diagnostics and therapeutics for domestic animals. In addition, animal experiments may also aim at improved productivity of livestock.

In addition, animal experiments are used for educational purposes in schools of medicine and veterinary medicine. Those who are to carry out animal experimentation must be given proper training.

Let us turn now to the issue of production and experimental use of genetically modified animals.

2. Genetically Modified Animals

Genetically modified animals are examples of genetically modified organisms. Other examples of genetically modified organisms are genetically modified bacteria and plants. Within the European Union, we find the following definition:

> "genetically modified organism (GMO)" means an organism, with the exception of human beings, in which the genetic material has been altered in a way that does not occur naturally by mating and/or natural recombination (Directive 2001/18/EC).

We see here that the distinguishing feature of genetically modified organisms is that their genetic make-up does not occur naturally. By implication, the modification is carried out by means of a special technology.

In a Royal Society Report, "genetically modified animals" are defined as

> animals modified either via a technique known as transgenesis (when individual genes from the same or a different species are inserted into another individual) or by the targeting of specific changes in individual genes or chromosomes within a single species—targeted removal of genes (knock-outs) or targeted addition of genes (knock-ins) (Royal Society, 2001, p. 3).

Here the focus is explicitly on technology. Genetically modified animals are the result of transgenesis or gene targeting.

Therefore, both the European Union definition of genetically modified organisms and the definition offered in the Royal Society Report suggest that the technology determines what should count as a genetically modified animal. Animals modified by conventional breeding or spontaneous mutations are excluded.

By far the most common species that has been genetically modified is the mouse. The second most common is the rat. Other examples are sheep, pig, poultry, and different kinds of fish (Pinkert, 2002). In 2001 the results from the first genetic modification of a rhesus monkey were published. The monkey was called "ANDi," the name being the reverse of "i(nserted) DNA." A gene producing a green fluorescent protein was inserted in order to test the method (Chan *et al.*, 2001).

Scientists put forward many different reasons for producing and using genetically modified animals in research. For example, genetically animals may be used:

(1) in studies of gene function;
(2) as disease models;
(3) for production of therapeutic proteins ("bioreactors," "biopharming");
(4) for xenotransplantation (genetic modification of animal organs for human transplants);
(5) in toxicity testing; and
(6) for improving farm animal health and productivity (Mepham *et al.*, 1998; Royal Society, 2001).

The use of genetically modified animals in research is increasing (Stokstad, 1999; Royal Society, 2001; Nuffield Council on Bioethics, 2005), but the number varies depending on the purpose. For natural reasons, it can be expected that relatively few animals are modified for use in xenotransplantation research and for use as bioreactors, while very high numbers are used as disease models and for discovering gene function. Producing genetically modified animals is difficult and expensive, so many scientists prefer to buy them from specialized companies—for example, the Jackson Laboratory—or obtain them from colleagues.

There exist many different methods for generating genetically modified animals. The most common are pronuclear microinjection (in the Royal Society report called "transgenesis") and the embryonic stem cell method (in the Royal Society report called "targeted removal" and "targeted addition"). In addition, conditional methods are becoming increasingly important (Houdebine, 2003; Pinkert, 2002; see Chapter Six for a brief presentation).

With these methods several different types of genetic modification can be generated. One type is the insertion of a new gene. This gene may come from human beings or from another species. The gene produces a new protein in the animal cells that has never been produced before. Another type is when genetic material is inserted that leads to overexpression, that is, to the production of a larger amount of a particular protein. A third type is "knock-out," that is, a gene is inactivated. Each type of genetic modification can in principle be conditional or non-conditional. Conditional modifications can be tissue-specific or temporally specific, that is, they can be expressed in particular tissues or can be turned on or off at particular points of time.

Animal cloning represents another type of genetic modification. Here it is not a matter of transferring a single gene into the genome of an animal, but of transferring the whole nuclear genome from a somatic cell of an animal into an enucleated oocyte from another (of the same species). In this way an animal is created that is identical to another (mitochondrial DNA excepted). This somatic cell nuclear transfer can be combined with single gene modifications in the nuclear genome before the nuclear transfer.

The well-known "Dolly the sheep" was the result of the first successful cloning of a mammal (Wilmut *et al.*, 1997). To date, animal cloning has been carried out on several other animal species, including cow, goat, pig, mouse, rat, rabbit, cat, mule, horse, and dog (Lee *et al.*, 2005). Recently, an embryo of a rhesus monkey was cloned but only in order to obtain embryonic stem cells (Byrne *et al.*, 2007).

There might be several different reasons why scientists carry out animal cloning. Here is a list of potential uses of this technology:

(1) production of pharmaceutical proteins in animal milk ("bioreactors");
(2) cloning of potential breeding animals such as farm animals or sport animals with special traits (although cloning is probably not good for breeding generally and in the long-term);
(3) replacement of deceased companion animals;
(4) saving endangered species (preservation of biological diversity);
(5) animal experimentation (in which as similar individuals as possible are desirable);
(6) basic biological knowledge;
(7) xenotransplantation (in combination with genetic modification); and
(8) medical application to human beings (therapeutic cloning in order to treat diseases).

Animal cloning is often combined with genetic modification. In the list this has already been pointed out regarding xenotransplantation. It is also the case with regard to bioreactors and sometimes also with regard to breeding animals. Human reproductive cloning is commonly not an objective for carrying out animal cloning experiments. On the contrary, human reproductive cloning is often explicitly rejected by scientists conducting this type of research (see, for example, Jaenisch and Wilmut, 2001). Human therapeutic cloning might still be an objective.

Finally, a comment on the terms "genetically modified animal" and "transgenic animal," starting with the latter. The term "transgenic animal" is sometimes used only for animals with a gene from another species, sometimes also for animals with a gene from the same species. Sometimes it is used only when genes have been transferred through pronuclear microinjection, now and then also for "knock-ins" by means of the emrbryonic stem cell method. Sometimes it is used also for "knock-outs," since DNA in this case is transferred in order to achieve the gene inactivation. Sometimes "transgenic" is used also for modifications carried out by conditional methods. In this book, I prefer the term "genetically modified animals" as an umbrella term for all these different types of modifications. This term is broader than some of the uses of the term "transgenic animals." The term "genetically modified animals" will also include cloned animals.

3. Public Attitudes toward Animal Experimentation

What, then, are the general public's views on animal experimentation? And what are its views on research involving genetically modified animals?

Perhaps the most enlightening survey to date regarding animal experimentation in general is the 1999 MORI (Market and Opinion Research International) poll from Great Britain (MORI, 1999). The poll was initiated by the journal *New Scientist* (Aldhous *et al.*, 1999). It is now fairly old but still very interesting.

The interview sample (n=2009 persons aged 15+) was divided into two groups. The first half was given a "cold start." They were asked whether they agreed or disagreed that scientists should be allowed to carry out animal experiments. The second half of the sample was given a "warm start." They were told: "Some scientists are developing and testing new drugs to reduce pain, or developing new treatments for life-threatening diseases such as leukemia and AIDS. By conducting experiments on live animals, scientists believe they can make more rapid progress than would otherwise have been possible" (MORI, 1999).

In the group that got a cold start, only 24% were in favor of animal experimentation, while 64% were against. In the group that got a warm start, on the other hand, 45% were for and 41% against. This represents a swing of 22% from disapproval to approval. This is a large swing for this kind of survey (MORI, 1999; Aldhous *et al.*, 1999).

These results show that the way in which the questions are posed may affect the answers. They indicate that many people seem prepared to change their views if they receive more information.

The MORI poll also investigated attitudes toward different types of animal experiments. A range of goals for animal experiments was selected and the participants were asked whether they approved or disapproved if: (a) animals do not suffer; (b) animals are subjected to pain, illness or surgery; and (c) animals may die. Again, the sample was divided. One-half was told that the experiments would be on mice, the other that they would be on monkeys.

The results show that a majority is prepared to accept that mice may suffer, if this helps in finding treatment for life-threatening diseases such as childhood leukemia or AIDS. If the experiment is on mice, will not cause suffering, and aims at developing or safety-testing a drug to treat leukemia, as many as 83% approve. Opinion was evenly divided regarding experiments to develop and test a painkilling drug if the experiment involved mice suffering pain. This means that many people found an ethically relevant difference between experiments targeting life-threatening diseases and those that do not.

Experiments on monkeys were viewed much more negatively than were experiments on mice. Only research aimed at finding treatment for childhood leukemia was seen as justifying the suffering of monkeys in experimentation.

Experiments investigating the sense of hearing met particular opposition. A large majority accepted the use of mice in hearing experiments if they were not harmed, but such experiments showed the biggest swing toward disapproval if the mice were subjected to pain, illness, or surgery.

Finally, people did not find experiments in which animals might die more questionable than those in which animals are subjected to pain, illness, or surgery (MORI, 1999; Aldhous *et al.*, 1999).

These findings indicate that approval and disapproval depend to a large extent on exactly which goal the experiment has and which animal species is experimented on. The participants carried out a case-by-case balancing of expected human benefits against animal suffering and the species involved. As suggested by the attitudes toward hearing experiments, basic research involving animal harm appears controversial compared to clinically oriented research. Scientists might find it difficult to convince people of its justification, although they might succeed regarding those experiments in which animal suffering is mild.

Attitudes toward animal experimentation may vary from one country to another. The MORI poll only concerns Britain, although Aldhous *et al.* state that many other countries would likely have expressed similar views. This may hold true, for instance, elsewhere in northern Europe where animal welfare organizations have a similarly high profile. Surveys in the United States suggest that Americans are more positive about animal experimentation than are the British (Aldhous *et al.*, 1999).

The quantitative results from the MORI survey can be compared with the results of the focus group study by Phil Macnaghten. In this study, there were eight groups, six of which had specific relations to animals and two without any such relations. Six groups came from the North-West of England: pet owners, wildlife enthusiasts, intensive farmers, extensive farmers, country sports enthusiasts, and a non-animals group, comprising a control group. In addition, there were two London groups, one consisting of pet owners and a non-animals group, functioning as a control. The focus group discussions indicated only limited appreciation in the United Kingdom of animal experimentation and testing. People's attitudes depend on the purpose of the research. They feel less uncomfortable about animal testing for medical purposes than for cosmetic ones (Macnaghten, 2001; Macnaghten, 2004). These results are quite in line with those of the MORI poll.

4. Public Attitudes toward Genetically Modified Animals

Several Eurobarometer surveys of public attitudes toward different types of biotechnology—initiated by the European Commission—have been carried out. The survey from 1996 is particularly interesting in our context, since it included a question about attitudes toward genetic modification of laboratory animals (Eurobarometer 46.1, 1997). This was not the case in the more recent survey from 1999, which focused instead on animal cloning (Eurobarometer 52.1, 2000). The survey from 2002 focused neither on genetic modification of laboratory animals nor on animal cloning. However, like the 1996 survey it included a question about xenotransplantation (Eurobarometer 58.0, 2003). A more recent Eurobarometer survey did not focus on attitudes toward animals at all (Eurobarometer 64.3, 2006).

In the 1996 Eurobarometer survey, the attitudes of the public to six different applications of gene technology were studied: food production, crop plants, bacteria for production of pharmaceuticals, laboratory animals, xenotransplants, and genetic testing. On the questions of genetically engineering animals for laboratory research and xenotransplantation, the majority of 16,000 people throughout the European Union felt that although these two types of genetic modification might be useful (57.7% and 51.0%, respectively), they are risky (51.7% and 58.8%), morally unacceptable (47.8% and 52.5%), and should not be encouraged (43.8% and 48.3%)(Eurobarometer 46.1, 1997; Durant *et al.* 1998, pp. 250–258). In the 1999 survey, attitudes toward animal cloning were investigated. The result showed that such cloning had quite low support (Eurobarometer 52.1, 2000).

In general, the Eurobarometer surveys indicated that complete rejection of biotechnology was rare among Europeans. Medical applications tended to be considered more acceptable than applications for food. Almost as unacceptable as food applications were xenotransplantation, genetic modification of laboratory animals, and animal cloning.

The correlation between knowledge and attitude was very low. Countries showing the highest levels of knowledge—for example, Sweden and Denmark—were the most opposed to genetic engineering in general (Durant *et al.*, 1998, pp. 199–200).

The Swedish psychologist Lennart Sjöberg carried out a survey of the attitudes of Swedes toward gene technology, which was methodologically more rigorous than the Eurobarometer surveys. As in the Eurobarometer surveys, Sjöberg found that medical applications were well accepted, while food applications were less so. In particular, there was a negative attitude toward genetic modification of animals for food purposes. Key factors behind these attitudes were that food applications were perceived as being of no real benefit and at the same time risky (Sjöberg, 2004, pp. 47–53).

In Macnaghten's focus group study in the United Kingdom mentioned above, the issue of genetically modified animals was also discussed. Most participants viewed genetic modification of animals as both "new" and "unnatural." Although few people were completely opposed to this technology, there was considerable concern about the speed and pace of its development, the degree of intervention and precision, and the likelihood of unexpected mistakes. The vital importance of demonstrating a genuine need for genetic modification was stressed. Public concerns included concerns about the intrinsic character of animals, including animal integrity, and concerns about animal welfare. The necessity of regulation and institutional oversight was also of strong concern (Macnaghten, 2001; Macnaghten, 2004).

Let me also refer to a focus group study carried out in Denmark (Lassen *et al.*, 2006). This study confirmed the results of the Eurobarometer surveys, namely that medical applications were assessed most positively and food applications most negatively. The investigators found several different categories of arguments for and against animal biotechnology. The *pro* arguments were caught in terms of utility such as economic usefulness, social usefulness, and self-interested usefulness. The *con* arguments referred to risk (environment, health) and utility (no need, wrong strategy). But there were also other arguments focusing on animal welfare and the integrity of nature.

The issue of scientific literacy and public understanding of science has been subject to much discussion (Miller, 1998). It is interesting to compare the results of the Eurobarometer with those of the MORI poll in this regard. The results of the latter indicate that the way in which we pose the questions affects the answers. They also indicate that portions of the general public seem prepared to change their views if they receive more information. This might suggest that if people were more "scientifically literate" they would be more positive to animal experimentation. The Eurobarometer, on the other hand, indicates that scientific literacy might not suffice; informed people might still say "no" for ethical reasons (*cf.* Lassen *et al.*, 2006). Taken together, the MORI poll and the Eurobarometer surveys seem to indicate that active supporters and active opponents of science and technology may be motivated to search for more scientific knowledge.

5. The Philosophical Debate

These polls, surveys, and focus group studies indicate that animal experimentation is a controversial issue from an ethical point of view. Animal experimentation in general and the production and experimental use of genetically modified animals in particular evoke strong feelings among the general public. In the academic literature, we see a similar spectrum of views, although these views are much more elaborately expressed and well-argued.

Many authors on animal ethics are—to a varying degree—critical of animal experimentation (for example, Singer, 1993a; Singer 1995; Regan, 1983; DeGrazia, 1996; LaFollette and Shanks, 1996; Greek and Greek, 2001; Greek and Greek, 2002). Some authors are very positive (for example, Carruthers, 1992; Cohen, 1994). Surprisingly few authors propose a middle course in between these extremes. It is surprising, since such a middle course—as indicated by the above polls, surveys, and focus group studies—appears quite common among the general public. In this book I try to develop such a middle course, and in doing so I am inspired by the British philosopher Mary Midgley (1983).

A special feature of the book is that I discuss the concept of animal welfare more than is common in philosophical books on animal ethics. The reason is that I believe that it is vital to bring animal ethics closer to animal welfare science. In this regard, I am inspired by David Fraser (1999). My contribution can be viewed as an attempt to combine Midgley's type of animal ethics with Fraser's conception of animal welfare.

6. Outline

The book consists of five chapters, apart from this first introductory chapter. In Chapter Two, I analyze five key positions in the ethics of animal experimentation, ranging from the most positive to the most negative. I do so by focusing on a clear example of each category called "a prototype."

In the first section of Chapter Three, the legal regulations of several important research countries are related to the five prototypes. Then a substantial part of the chapter is devoted to an analysis of basic presumptions of each of the five prototypes. I also criticize the prototypes and propose a version of the "weak human priority" view. I argue that although animal experimentation inflicting pain is *prima facie* wrong, it can be accepted given certain special considerations. I put forward an argument from species care according to which the special obligations to our children and other human beings to find medical treatments commonly—but not always—outweigh our obligations to animals. Some possible animal experiments are not acceptable, since the expected human benefit is too low and the animal suffering too severe. At the end of the chapter, I present the ethical theory that lies at the foundation of this position in the ethics of animal experimentation. In subsequent chapters I develop other aspects of this position such as the scientific value of animal experimentation, the concept of animal welfare, and ethical balancing.

Chapter Four focuses on the scientific value of animal experimentation. The standard *pro* argument for its scientific value involves giving examples from the history of medicine of valuable experiments. The most important *con* argument, on the other hand, stresses causal disanalogy, that is, that causal mechanisms may not be similar in animal models and in human beings, or—put in other words—that data obtained from animal models cannot be reliably extrapolated to human beings. Several objections to this *con* argument are discussed. I argue that it is necessary in many cases to use intact animals, but I also acknowledge some problems of extrapolating from animals to human beings. The "3Rs" of Russell and Burch (1992) are also discussed: replacement, reduction and refinement.

In Chapter Five, three different types of animal welfare conceptions are discussed: function-based, feeling-based, and those focusing on natural living. In practice these conceptions often overlap but sometimes they conflict. I suggest a comprehensive approach including all three concerns—inspired by David Fraser—but argue that in animal experimentation the aspect of feeling is the most important. Animal sentience is therefore discussed in more detail. A practical "checklist" is given regarding animal welfare aspects of an animal experiment, related to pre-procedural concerns, experimental concerns, and post-procedural concerns. In the final part of the chapter I discuss the nature of ethical balancing and analyze different models of ethical balancing in relation to animal experimentation.

The last chapter (Chapter Six) includes an analysis of the implications of the five prototypes of animal experimentation for the production and experimental use of genetically modified animals. Three main concerns are discussed in relation to genetically modified animals: scientific concerns, intrinsic ethical concerns, and animal welfare concerns. Finally, I present and discuss four cases.

Two

FIVE ETHICAL PROTOTYPES OF ANIMAL EXPERIMENTATION

Many different ethical views on animal experimentation have been proposed. These views can be categorized in different ways. I will analyze ethical views belonging to five different categories: human dominion, equal consideration of interests, animal rights, strong human priority, and weak human priority. I will do so by focusing on a clear example within each category. Such a clear example can be called a "prototype." The prototypes are surrounded with non-prototypical views. These non-prototypical views are unclear examples in the gray area between different categories of views. The five ethical prototypes of animal experimentation to be analyzed here are as follows (for a quite similar categorization, see Brody, 1998, pp. 15–18; for a different categorization, see Orlans, 1993, pp. 20–34):

(1) human dominion (proposed by Peter Carruthers);
(2) equal consideration of interests (Peter Singer);
(3) animal rights (Tom Regan);
(4) strong human priority (Carl Cohen); and
(5) weak human priority (Mary Midgley).

Let me clarify the order of presentation. I have chosen to combine a historical and a theoretical perspective. Prototype (1) belongs to a category of views that has played an important role historically. This category constitutes a type of position that the other views—one way or another—react against. For this reason, it is warranted to start the presentation with this category. The prototypical example as such is fairly recent, however. In order to clarify the background to this prototype, I add a section providing historical perspectives. Prototype (2) started to a large extent the present-day philosophical debate on animal ethics and is a radical defense of animal interests. Prototype (3) is even more radical, defending animal rights. Prototype (4) is a strong reaction to both prototype (2) and prototype (3), promoting human research interests. Prototype (5) represents a middle course.

Thus, the order of presentation is not from one extreme to the other. Such a presentation—from the most positive to the most negative ethical view of animal experimentation—would instead look like this: human dominion, strong human priority, weak human priority, equal consideration of interests, and animal rights.

In animal ethics and the ethics of animal experimentation, many other positions than those presented here have been suggested (see, for example, Frey, 1980; Rowan, 1984; Sapontzis, 1987; Rollin, 1989; Rodd, 1990; Orlans,

1993; Linzey, 1995; DeGrazia, 1996; LaFollette and Shanks, 1996; Greek and Greek, 2001; Greek and Greek, 2002; see also Nuffield Council on Bioethics, 2005). However, the selected views are fairly representative of the modern debate. I will discuss some aspects of other positions in other chapters (DeGrazia in several different chapters, LaFollette and Shanks in Chapter Four, and Rollin in Chapter Five), but a more extensive analysis of other views is beyond the purpose of the book. I will not discuss a purely relational animal ethics, but I will analyze in detail Midgley's mix of a relational approach and interspecies justice. Not all selected authors discuss the ethics of animal experimentation extensively. The reason I nevertheless focus on these authors is that they have been very influential in the general animal ethics debate, and that their views have clear and easily recognized implications for the ethics of animal experimentation.

Let me stress that the analysis has the character of an overview. A deeper comparative analysis of some important aspects will be carried out in the next chapter, although in order to clarify the views I now and then include some comparisons already in this chapter. Criticism of the prototypes in terms of an assessment of their respective strengths and weaknesses will also be left to the next chapter.

1. Human Dominion

The basic idea of the human dominion category of views is that all animal experimentation that can be expected to benefit human beings is ethically acceptable. A key defender of this view in the present-day discussion is Peter Carruthers (1992).

A. Contractualism

Carruthers's ethical theory is contractualist and puts human beings at the center of ethical concern. Morality is constructed by human beings in order to facilitate cooperation within the community. Human beings agree to reciprocal rights and duties (Carruthers, 1992, p. 102). The contract is not a historical event but a metaphor for morality as a matter of common agreement. Morality is not discovered in nature—as in the tradition of natural rights—but invented by human beings. Carruthers indicates the pragmatic function of morality. It facilitates human cooperation. The key aspect of Carruthers's view of morality is reciprocity. Rights and duties are two sides of the same coin.

B. Implications for Animal Ethics

With regard to animals, Carruthers argues as follows:

> On this approach animals, like buildings, would have no direct rights or moral standing. Rather, causing suffering to an animal would violate the

right of animal lovers to have their concerns respected and taken seriously. Such an approach may be able to recover for contractualism a great deal of what common-sense tells us about the moral treatment of animals. In particular, it can explain how it can be true that, while we do have duties towards animals, their lives and interests cannot be weighed against the lives and interests of humans. For the duties in question only arise indirectly, out of respect for those who care about animals (Carruthers, 1992, pp. 106–107).

We see here that—according to Carruthers—we have no direct duties toward animals, only indirect ones. This means that we should not treat animals well for their own sake but for the sake of other human beings with whom we have entered a moral contract. We have direct duties to respect the feelings of animal lovers, and we should treat animals well for their sake. We have only indirect duties to animals, and animals have correspondingly only indirect rights.

Carruthers explicitly states that animals lack moral standing. This is a quite extreme view in the modern debate. Even those who rank human beings higher than animals still assign animals some moral standing.

Carruthers also criticizes explicitly the idea of balancing human interests and animal interests, an idea that is central both to the prototype of equal consideration of interests and the weak human priority prototype. The interests of animals count for virtually nothing compared to human interests.

These two ideas—the idea that animals lack moral standing and the idea that balancing human interests and animal interests has no place—are key ideas in Carruthers's ethical view and a reason why I call it a "human dominion" prototype. Human beings are free to use animals for their own benefit, for example, in scientific research. The only restrictions are due to other human beings.

C. The Difference between Human Beings and Animals

In arguing for his position, Carruthers stresses that a sharp distinction exists between human beings and animals, but not within the human species. He maintains that

> there are no sharp boundaries between a baby and an adult, between a not very intelligent adult and a severe mental defective, or between a normal old person and someone who is severely senile... In contrast, there really is a sharp boundary between human beings and all other animals. Not necessarily in terms of intelligence or degree of rational agency, of course—a chimpanzee may be more intelligent than a mentally defective human, and a dolphin may be a rational agent to a higher degree than a human baby (Carruthers, 1992, pp. 114–115).

The sharp distinction between human beings and animals stressed in this quotation should be compared with the analogy between animals and buildings pointed out in the previous one above. This analogy indicates that Carruthers—from a moral perspective—accepts only two basic kinds of being, namely being a rational agent and being a thing. To be an animal is to be a thing. Animals lack moral standing.

The reason animals lack moral standing is that they are not rational agents "in the sense necessary to secure them direct rights under contractualism" (Carruthers, 1992, p. 145). Carruthers does not rule out some animals' having some rational ability. The point is that they do not have it in a sufficient degree.

D. Animals Do Not Feel Pain

In a more speculative chapter of his book Carruthers maintains that not only are animals not rational agents in the full human sense, but they are unconscious. He suggests that "human beings are unique amongst members of the animal kingdom in possessing conscious mental states" (Carruthers, 1992, p. 186). Given the present-day academic debate, this is a quite extreme, neo-Cartesian view (*cf.* DeGrazia, 1996, pp. 53–56). Carruthers explains:

> It seems that pain, like all other mental states, admits of both conscious and non-conscious varieties… If animals are incapable of thinking about their own acts of thinking, then their pains must all be non-conscious ones (Carruthers, 1992, p. 189).

On the basis of this belief that animals are non-conscious, Carruthers makes the ethical claim that animals have no moral standing. He argues: "For if it has been shown that the mental states of animals are non-conscious, then they cannot be appropriate objects of moral concern" (Carruthers, 1992, p. 192).

Carruthers's claims are only tentative:

> I would urge caution, however. The views presented in this chapter are controversial and speculative, and may well turn out to be mistaken. Until something like consensus emerges…, it may be wiser to continue to respond to animals as if their mental states were conscious ones (Carruthers, 1992, pp. 192–193).

This tentativeness is not acknowledged sufficiently by David DeGrazia. He appears to interpret Carruthers more strictly (DeGrazia, 1996, pp. 53–56, 112–115). In Chapter Five, I will discuss the issue of animal sentience in more detail. I will there accept DeGrazia's argument that we have strong scientific reason to believe that many animals are sentient and can feel pain. By stating that "if animals are incapable of thinking about their own acts of thinking, then their pains must all be non-conscious ones" (Carruthers, 1992, p. 189),

Carruthers appears to conflate the basic ability of having unpleasant feelings and thinking, which is something much more advanced and intellectual.

In its non-speculative version Carruthers's argument is more adequate than in its speculative one with regard to empirical aspects: many animals can feel pain. However, it is weaker normatively, because if animals can feel pain, why would this not be ethically relevant? Even if we embrace a contractualist theory, there would be reason to supplement this view with some kind of view that takes animal pain seriously. And animal pain could be considered ethically relevant even if we maintain that human interests outweigh this pain. But we can easily understand that Carruthers wants to make his ethical argument stronger by basing it on a radical skepticism regarding the ability of animals to feel pain.

E. Animal Experimentation

Carruthers only makes a few comments on animal experimentation, but his view on animal ethics in general is straightforward, and it is fairly easy to recognize its implications for the ethics of animal experimentation. Since animal experimentation can be expected to reduce human suffering it is completely acceptable. No conscious animal pain exists that can outweigh this benefit, and if some such pain exists—remember that Carruthers's proposal is tentative—it does not outweigh it anyhow.

Carruthers discusses explicitly only one particular aspect of animal experimentation, namely the effects on human character of carrying out such experimentation. He argues:

> For example, consider technicians working in laboratories that use animals for the testing of detergents, causing them much suffering in the process. That they can become desensitised to animal suffering in such a context provides little reason for thinking that they will be any less sympathetic and generous outside it (Carruthers, 1992, p. 159).

Thus, Carruthers argues that becoming desensitized in one context does not necessarily mean that one becomes desensitized in another. This can be compared to Kant's view that acting brutally toward animals may dehumanize us. In contrast to Kant, Carruthers downplays the negative effects on human character of inflicting pain on animals.

F. Conclusion

In sum, the human dominion prototype proposed by Carruthers implies that animal experimentation is ethically acceptable if it can be expected to lead to human benefit. We should care for animals used in research but not for their sake, only for the sake of other human beings who care for animals.

G. Historical Perspectives

The human dominion view—or category of views—is rare in the present-day academic debate on the ethics of animal experimentation, although it might be more common among scientists and the public, at least in weaker, non-prototypical, versions. In earlier centuries, however, many prominent thinkers defended this view.

In order to put Carruthers's view into perspective, let me give a brief historical overview of earlier examples of this category of views, starting with the Bible. In Genesis 1:26 (New Revised Standard Version), we read:

> Then God said, "Let us make humankind in our image, according to our likeness; and let them have dominion over the fish of the sea, and over the birds of the air, and over the cattle, and over all the wild animals of the earth, and over every creeping thing that creeps upon the earth."

We see here the metaphor of human "dominion" at the Judeo-Christian root of Western culture (the Ancient Greek root of Western culture will not bother us here). Humankind is created in the image of God and differs radically from animals. God gives humankind dominion over the animals.

The famous Christian theologian Augustine explains—around 400 AD—the difference between human beings and animals in the following way:

> If, when we say, Thou shalt not kill, we do not understand this of the plants, since they have no sensation, nor of the irrational animals that fly, swim, walk, or creep, since they are dissociated from us by their want [lack] of reason, and therefore by the just appointment of the Creator subjected to us to kill or keep alive for our own uses; if so, then it remains that we understand that commandment simply of man. The commandment is "Thou shalt not kill man" (Augustine, 1877, Book 1, pp. 31–32).

Animals differ from human beings by lacking reason. This fact—together with the will of God—justifies the killing of animals by human beings, and their use of them for their own benefit. Modern theologians commonly stress that human beings are to be the stewards of Creation instead of dictators (and a radical author like Andrew Linzey, in his book *Animal Theology*, goes even a step further and stresses the role of human beings as liberators; throughout history human beings have enslaved animals and now they have an obligation to put an end to that slavery (Linzey, 1995)). It is still obvious, historically speaking, that human beings are generally viewed in the Christian tradition as being in some kind of position of dominion over animals.

The difference between human beings and animals has never been stressed more strongly than by René Descartes, the 17th century philosopher. In a sophisticated argument, he makes an analogy between animals and machines:

> Here I specially stopped to show that if there had been such machines, possessing the organs and outward form of a monkey or some other animal without reason, we should not have had any means of ascertaining that they were not of the same nature as those animals. On the other hand, if there were machines which bore a resemblance to our body and imitated our actions as far as it was morally possible to do so, we should always have two very certain tests by which to recognise that, for all that, they were not real men. The first is, that they could never use speech or other signs as we do when placing our thoughts on record for the benefit of others ... And the second difference is, that although machines can perform certain things as well as or perhaps better than any of us can do, they infallibly fall short in others, by the which means we may discover that they did not act from knowledge, but only from the disposition of their organs ... And this does not merely show that the brutes have less reason than men, but that they have none at all (Descartes, 1997, pp. 107–108).

Descartes argues that it would not be possible to distinguish an animal from a machine but that it would be possible to distinguish a human being from a machine. In distinction to human beings, animals cannot use a language and they never act on the basis of knowledge. Animals lack consciousness. It is not the case that animals have a lower degree of consciousness; they lack it completely. Descartes assumes an all-or-nothing view of consciousness (*cf.* Hursthouse, 2000, pp. 65–69).

The 18th century philosopher Immanuel Kant argues that we have direct duties only to human beings, although we have indirect duties to animals. He states:

> But so far as animals are concerned, we have no direct duties. Animals are not self-conscious and are there merely as means to an end. That end is man. We can ask, "Why do animals exist?" But to ask, "Why does man exist?" is a meaningless question. Our duties towards animals are merely indirect duties ... If a man shoots his dog because the animal is no longer capable of service, he does not fail in this duty to the dog, for the dog cannot judge, but his act is inhuman and damages in himself that humanity which it is his duty to show towards mankind (Kant, 1963, pp. 239–240).

According to Kant, we cannot have direct duties to animals because they are not self-conscious. They exist only as means to an end. They are not ends in themselves, as human beings are. Our duties to animals are therefore only indirect. They are ultimately duties to human beings. Acting brutally toward animals affects human character negatively. It reduces our humanity (*cf.* Carruthers, 1992, pp. 157–158).

In Western history, many other theological and philosophical views on animals have been held, views that do not belong to the category of human dominion. This may also be true of the views of ordinary people. It is, however, beyond the purpose of this presentation to investigate and discuss these views.

So far we have looked at human dominion views on animals in general. Let us turn to animal experimentation. Claude Bernard, the founding father of physiology and a devoted animal experimenter, writes at the end of the 19th century:

> Have we the right to make experiments on animals and vivisect them? As for me, I think that right, wholly and absolutely. It would be strange indeed if we recognize man's right to make use of animals in every walk of life, for domestic service, for food, and then forbade him to make use of them for his own instruction in one of the sciences most useful to humanity. No hesitation is possible: the science of life can be established only through experiment, and we can save living beings from death only after sacrificing others... It is essentially moral to make experiments on an animal, even though painful and dangerous to him, if they may be useful to man (Bernard 1949, pp. 100–101).

In this quotation, Bernard gives three arguments for animal experimentation that are still key arguments in the modern discussion. First, if we accept other uses of animals and want to be consistent, we should also accept the use of animals in scientific experimentation. Second, animal experimentation is necessary for "the science of life." Third, we can save human lives only if we accept that animals are sacrificed in experiments.

In an attempt to defend himself from the accusation of being brutal, Bernard says:

> The physiologist is not an ordinary man: he is a scientist, possessed and absorbed by the scientific idea he pursues. He does not hear the cries of animals, he does not see their flowing blood, he sees nothing but his idea, and is aware of nothing but an organism that conceals from him the problem he is seeking to resolve (Bernard 1949, p. 102).

In this psychological defense, Bernard presents the image of the devoted scientist burning for a scientific idea and allowing nothing to stop him (*cf.* Orlans, 1993, p. 15). By implication the Bernardian scientist does not care about animal suffering at all. He does not try to minimize animal pain. Human research interests make animal suffering completely irrelevant.

2. Equal Consideration of Interests

We have seen that many historically important proponents of the human dominion view have embraced the idea that the ethically relevant characteristic is consciousness or rationality in the full human sense. However, not all philosophers of earlier times shared this view. A counter-example is Jeremy Bentham. He criticizes explicitly the view that reason or linguistic ability are ethically relevant and argues that the ethically relevant property on the contrary is the ability to suffer:

> A full grown horse or dog is beyond comparison a more rational, as well as more conversable animal, than an infant of a day, or a week or even a month old. But suppose the cause were otherwise, what would it avail? The question is not, Can they *reason*? nor Can they *talk*? but, *Can they suffer*? (Bentham, 1789, Chapter XVIII)

This brings us to the second prototype of animal experimentation, namely "equal consideration of interests."

A modern philosopher who is influenced by Bentham and takes the ability to suffer seriously is Peter Singer. His book *Animal Liberation* from 1975 became extremely influential, not only among the general public but also among philosophers. In a way, it started the modern debate on animal ethics, or at least contributed to this start in a substantial way. His ideas on animal ethics and other topics in applied ethics were also developed in *Practical Ethics* from 1979. I will analyze Singer's view on animal experimentation as it emerges in the second editions of these two books (Singer, 1995; Singer, 1993a).

A. Radical Criticism of Animal Experimentation

Singer is very critical of animal experiments, although he does not reject them entirely. The more precise extent to which he accepts them is difficult to determine. He never states it explicitly. In a sense, the question of the extent to which animal experiments are acceptable is not meaningful. It is not possible to indicate a particular percentage, but it is possible to determine whether it is Singer's view that many (the weak interpretation) or most (the strong interpretation) animal experiments, as they are carried out in present-day research, are ethically unacceptable.

At first sight, Singer merely suggests that many animal experiments are unacceptable, indicating that the weak interpretation is the most adequate one. He starts out criticizing the notion that all animal experiments are beneficial:

> People sometimes think that all animal experiments serve vital medical purposes, and can be justified on the grounds that they relieve more suffering than they cause. This comfortable belief is mistaken (Singer, 1993a, p. 65).

Giving a few examples, he concludes:

> In these cases, and many others like them, the benefits to humans are either non-existent or uncertain; while the losses to members of other species are certain and real (Singer, 1993a, p. 66).

Singer plays out the uncertainty of human benefits against the certainty of animal suffering. The quotation is compatible with the weak interpretation that many but not most animal experiments are ethically unacceptable.

However, in *Animal Liberation* a statement appears that is in line with the strong interpretation, that is, that most animal experiments are ethically unacceptable:

> Among the tens of millions of experiments performed, only a few can possibly be regarded as contributing to important medical research (Singer, 1995, p. 40).

This statement that "among tens of millions ... only a few" have important medical benefits indicates that the strong interpretation is the adequate one.

Singer is particularly critical of the use of animal experiments in basic research:

> The broad label "medical research" can also be used to cover research that is motivated by a general intellectual curiosity. Such curiosity may be acceptable as part of a basic research for knowledge when it involves no suffering, but should not be tolerated if it causes pain (Singer, 1995, p. 61).

But even if it is a matter of a crucial medical experiment, it is not automatically acceptable. Singer is much more cautious. He makes the merely hypothetical case:

> If one, or even a dozen animals had to suffer experiments in order to save thousands, I would think it right and in accordance with equal consideration of interests that they should do so (Singer, 1993a, p. 67).

He stresses, however, that in reality experiments do not have these dramatic results. We saw above that only a few animal experiments out of "tens of millions" have important medical benefits.

Singer's extremely restrictive view is underlined by the requirement that the experiment must be acceptable to perform also on brain-damaged human beings:

> So whenever experimenters claim that their experiments are important enough to justify the use of animals, we should ask them whether they would be prepared to use a brain-damaged human being at a similar mental level to the animals they are planning to use. I cannot imagine that anyone would seriously propose carrying out the experiments described in this chapter on brain-damaged human beings (Singer, 1995, p. 83).

He appears to view this requirement as a useful rule of thumb. He continues:

> An experiment [using animal subjects] cannot be justifiable unless the experiment is so important that the use of a brain-damaged human being would also be justifiable. This is not an absolutist principle. I do not believe that it could never be justifiable to experiment on a brain-damaged human. If it really were possible to save several lives by an experiment that would take just one life, and there were no other way those lives could be saved, it would be right to do the experiment. But this would be an extremely rare case (Singer, 1995, p. 85).

If we apply Singer's rule of thumb, what would be the result? It would be that animal experiments are ethically acceptable only in "extremely rare" cases. This supports the strong interpretation.

Singer's view on animal experiments can be summarized as follows:

(1) In many experiments the benefits to human beings are either nonexistent or very uncertain, while the costs to animals are very real.
(2) Only a few medical experiments out of tens of millions are important.
(3) Basic research involving animal pain is unacceptable.
(4) Hypothetically, a medical experiment involving animal pain would be acceptable, if merely a few animals were to be used and the experiment could be expected to save the lives of thousands of human beings.
(5) A medical experiment involving animal pain is acceptable only if the use of a retarded human being would also be acceptable, and this would be acceptable only in extremely rare cases.

My conclusion is that in Singer's view *most* experiments involving animal pain are ethically unacceptable. This means that the strong interpretation is the adequate one. Note, however, that the term "most" is my analytic term. I use it to understand Singer. It is not a term that he uses. As I use the term "most" in this context it means "at least a clear majority of animal experiments that are carried out in today's research."

This strong interpretation contradicts the analysis carried out by Rosalind Hursthouse. She interprets Singer as arguing for the weak interpretation that many—rather than most—animal experiments are unacceptable. This is due to the principle of charity, which in her view should govern all interpretation (Hursthouse, 2000, pp. 55–56). As far as I can see, this interpretation is too "charitable." It does not take seriously enough what Singer actually writes.

B. Impartiality and Its Implications

The starting point of Singer's argument for his view on animal experimentation is his understanding of what ethics is all about:

> An ethical principle cannot be justified in relation to any partial or sectional group. Ethics takes a universal point of view. This does not mean that a particular ethical judgment must be universally applicable. Circumstances alter causes... What it does mean is that in making ethical judgments we go beyond our own likes and dislikes (Singer, 1993a, pp. 11–12).

Central to his view is impartiality. The key characteristic of ethics is to go beyond our interests and the interests of our group, and assume a universal perspective. This view of ethics determines his views on all ethical issues, including animal ethics and the ethics of animal experimentation. From the very start, it excludes being partial with regard to human beings. This is vital to point out, because it may be thought that his view on animal ethics is determined by his specific normative ethical theory, namely utilitarianism. This is true in the sense that he applies his utilitarian theory to particular issues regarding the treatment of animals. The main reason, however, lies deeper, namely in his more fundamental presumptions about what ethics is all about. This means that if we are to criticize Singer regarding his view on animal experimentation, this is a first possible point to challenge.

The next step for Singer is to use impartiality as an argument for utilitarianism:

> We very swiftly arrive at an initially utilitarian position once we apply the universal aspect of ethics to simple, pre-ethical decision making. This, I believe, places the onus of proof on those who seek to go beyond utilitarianism (Singer, 1993a, pp. 12–14).

Singer here maintains without further argument that once we accept the "universal" perspective of ethics, we "very swiftly arrive" in utilitarianism. It is a little odd that Singer, who otherwise is very careful in his arguments, so "swiftly" comes to this conclusion. Many other ethical theories presume an impartial point of view, for example Kantian theories. But Singer is right that utilitarianism is compatible with this impartiality.

C. Equal Consideration of Interests

Singer's particular version of utilitarianism is interest utilitarianism (or preference utilitarianism). Its key idea is the principle of equal consideration of interests. This principle is described as follows:

> The essence of the principle of equal consideration of interests is that we give equal weight in our moral deliberations to the like interests of all those affected by our actions.... What the principle really amounts to is: an interest is an interest, whoever's interest it may be (Singer, 1993a, p. 21).

Who has a particular interest is ethically irrelevant. Only the *interest* as such is relevant. If the interest is well-being and avoidance of suffering, it does not matter whether that is the interest of a human being or a mouse.

D. Sentient Beings and Persons

In Singer's view, only sentient beings have interests. He argues:

> If a being is not capable of suffering, or of experiencing enjoyment or happiness, there is nothing to be taken into account. This is why the limit of sentience (using the term as a convenient, if not strictly accurate, shorthand for the capacity to suffer or experience enjoyment or happiness) is the only defensible boundary of concern for the interests of others (Singer, 1993a, p. 57–58).

Singer admits that sentience is perhaps not the best term to use, but nevertheless he uses it. Sentience refers to two types of experiential capabilities. One type is negative and refers to the ability to feel pain or suffer. The other is positive and represents the capacity for enjoyment or happiness. Singer means that the key interests of all sentient beings are related to avoiding negative feelings and having positive feelings.

It is not quite clear exactly which species are sentient in Singer's view (I will discuss this issue in Chapter Five), but it is a common view that all vertebrates are sentient and perhaps some invertebrates such as cephalopods.

Singer finds an ethically relevant difference within the set of sentient beings. He distinguishes sentient beings that are persons from those sentient beings that are non-persons. A "person" is "a rational self-conscious being" (1993a, p. 87).

Singer exemplifies the ethical relevance of this distinction with its implications for killing:

> For preference utilitarians, taking the life of a person will normally be worse than taking the life of some other being, since persons are highly future-oriented in their preferences (Singer, 1993a, p. 95).

Here we see that the key reason for not killing persons is that they have future-oriented preferences and killing would prevent them realizing these preferences.

Some members of other species are persons in the above sense, while some members of our species are not. Adult normal chimpanzees, for example, are persons in this sense, while newborns and some severely mentally retarded human beings are not (Singer, 1993a, p. 117).

Singer gives some examples of animal species in which adult normal individuals are persons and which it would be ethically unacceptable to kill:

> In the present state of our knowledge, this strong case against killing can be invoked most categorically against the slaughter of chimpanzees, gorillas and orangutans... A case can also be made, though with varying degrees of confidence, on behalf of whales, dolphins, monkeys, dogs, cats, pigs, seals, bears, cattle, sheep, and so on, perhaps even to the point at which it may include all mammals ... (Singer, 1993a, p. 132)

Singer expresses some hesitance about whether to include all mammals. If all adult normal mammals are viewed as persons, this would include also mice and rats, the most common species used in research. This would be controversial.

Killing sentient non-persons might, under some conditions, not be wrong:

> Thus it is possible to regard non-self-conscious animals as interchangeable with each other in a way that self-conscious beings are not. This means that in some circumstances—when animals lead pleasant lives, are killed painlessly, their deaths do not cause suffering to other animals, and the killing of one animal makes possible its replacement by another who would not otherwise have lived—the killing of non-self-conscious animals may not be wrong (Singer, 1993a, p. 133).

The reason for the condition that the animal has to be replaced is the utilitarian idea that otherwise the total happiness in the universe would be reduced. The happiness of the second animal compensates the loss of happiness caused by the killing of the first animal.

Singer does not develop the implications of the distinction between persons and non-persons for animal experimentation. Given his premises, it is worse to experiment on "persons" than on sentient animals that are not "persons," if the experimentation involves killing. However, even if mice and rats are not included in the set of persons, they still belong to the set of sentient beings, and inflicting pain on them in experiments would be ethically unacceptable, provided that the human benefit would not outweigh the animal pain, which is extremely rare.

E. Antispeciesism

A key term of Singer's is "speciesism." He explicitly says that he owes the term to Richard Ryder (Singer, 1995, p. 269). Singer defines "speciesists" as follows:

> Similarly those I would call "speciesists" give greater weight to the interests of members of their own species when there is a clash between their interests and the interests of those of other species. Human speciesists do not accept that pain is as bad when it is felt by pigs or mice as when it is felt by humans (Singer, 1993a, p. 58).

This definition has several special features. First, it has a focus on interests and pain. Second, the definition stresses that speciesism is at hand when a clash of interests occurs. Third, weighing interests is considered central to ethical decision-making. Fourth, speciesism is a matter of giving greater weight to human interests merely because of biological kinship, that is, that we belong to the same species. Taken together these features indicate that Singer's definition is strongly dependent on utilitarian presuppositions. Speciesism in Singer's sense is normative speciesism in distinction to descriptive speciesism. It is a view that we should give greater weight to human interests because we belong to the same species. Descriptive speciesism is the view that people embrace such a normative view or act in line with such a view. Singer recognizes that descriptive speciesism is at least partly true, that is, that many people embrace normative speciesism or act in line with it. Singer's assessment of normative speciesism is negative. Normatively, he is an antispeciesist.

Singer's definition of speciesism is unclear in some respects. It is unclear whether a person is a speciestist only at the moment when a clash of interests occurs or also at other points of time when no clash occurs but the person has the disposition to give greater weight to human interests in case of such a clash.

It is not quite clear whether speciesism is an all-or-nothing matter. Singer appears to use the term in the sense of always giving greater weight to human interests when a clash of interests occurs. However, we could imagine speciesism as a matter of degree. As we will see below, Midgley supports a moderate normative speciesism, according to which speciesism is sometimes justifiable and sometimes not.

Other possible definitions do not focus on interests but on obligations. Speciesism could, for example be defined as giving greater weight to obligations to human beings when a clash of obligations occurs. An advantage of such a definition is that it is less dependent on utilitarian presuppositions. It can also be accepted from a deontological point of view.

Singer stresses that a rejection of speciesism does not exclude ranking the value of different lives in some hierarchical ordering (Singer, 1993a, p.

107). It may be worse inflicting pain on normal adult human beings than on animals. Singer explains why:

> There are many areas in which the superior mental powers of normal adult humans make a difference: anticipation, more detailed memory, greater knowledge of what is happening, and so on. These differences explain why a human dying from cancer is likely to suffer more than a mouse (Singer, 1993a, p. 60).

But he continues:

> Yet these differences do not all point to greater suffering on the part of the normal human being. Sometimes animals may suffer more because of their more limited understanding (Singer, 1993a, p. 60).

These are important clarifications. Human beings might suffer more than non-human animals because of their higher mental capacities. But it might also be the other way around that non-human animals suffer more because of their lower mental abilities.

A useful test for the implications of Singer's normative antispeciesism is whether it would imply that if it comes to saving an adult normal chimpanzee from a burning house or a severely mentally retarded child, the chimpanzee should be saved. As Hursthouse points out, this appears to be a bullet that Singer is prepared to bite, despite that fact that it would be very controversial (Hursthouse, 2000, p. 119).

Finally, one more clarification. Singer's normative principle involves the equal consideration of interests, although our analysis has shown that only when the interests are equal should they be considered equally, not necessarily if the interests are different in some sense. Human beings may suffer more from a particular stimulus than non-human animals, but also less. With this in mind, Singer's principle should perhaps be rephrased in terms of equal consideration of equal interests.

F. Conclusion

In conclusion, the key ethical principle of Singer—and the term that I have used in designating his prototype—is equal consideration of interests. The application of this utilitarian principle leads him to condemn most experimentation involving animal pain that is carried out in today's scientific research. However, he is not against all such experimentation. A medical experiment involving animal pain could be ethically acceptable if merely a few animals were to be used and the experiment could be expected to save the lives of thousands of human beings. Thus, Singer should not be viewed as an abolitionist.

3. Animal Rights

Tom Regan is more radical than Singer. He is an abolitionist. He is against all animal experimentation involving animal harm. In the following analysis I will focus on Regan's main book *The Case for Animal Rights* from 1983. In this book he criticizes several other views in animal ethics such as indirect duty views and Singer's utilitarian view. As an alternative he develops a theory of animal rights. Regan applies this view to, for example, animal experimentation.

A. Total Elimination of Animal Experimentation

Regan summarizes his view on animal experimentation as follows:

> If we are seriously to challenge the use of animals in research, we must challenge the *practice* itself, not only individual instances of it or merely the liabilities in its present methodology. The rights view issues such a challenge. Routine use of animals in research assumes that their value is reducible to their possible utility relative to the interests of others. The rights view rejects this view of animals and their value, as it rejects the justice of institutions that treat them as renewable resources... Scientific research, when it involves routinely harming animals in the name of possible "human and humane benefits," violates this requirement of respectful treatment. Animals are not to be treated as mere receptacles or as renewable resources. Thus does the practice of scientific research on animals violate their rights. Thus ought it to cease, according to the rights view. It is not enough first conscientiously to look for nonanimal alternatives and then, having failed to find any, to resort to using animals. Though that approach is laudable as far as it goes, and though taking it would mark significant progress, it does not go far enough (Regan, 1983, pp. 384–385).

Several things are to be noted here. Regan refers to "the practice" of animal experimentation as a whole, not merely to individual experiments. More precisely, he talks about scientific research that involves "routinely harming" animals. It is stated that "harming" means failing to respect animals as individuals by treating them as resources to be used for human benefit. This violates their rights. Regan describes his view as "the rights view." Merely looking for alternative methods and then continuing to use animals when no such alternatives are found is not enough. Animal experimentation should be eliminated.

B. Rights Based on Inherent Value

Let us investigate Regan's "rights view" in more detail. Regan focuses on moral rights in distinction to legal rights. What are moral rights? A common definition that is also used by Regan is that moral rights are valid claims that have correlative duties (Regan, 1983, pp. 266, 271–273). But how are rights as claims validated, that is, established?

Regan stresses that rights are basic in the sense of being "unacquired." They are not dependent on a contract or a promise (Regan, 1983, p. 283). This can be compared to Carruthers's view. Carruthers's contractualist rights are acquired. They are the result of an agreement in society. Neither are the unacquired rights based on natural law, although Regan does not reject this idea explicitly. According to Regan, the unacquired rights are based on value, or—more precisely—on inherent value. He clarifies this in the following manner:

> Individuals who have inherent value have an equal basic right to be treated with respect. According to the rights view, this is a right that we can never be justified in ignoring or overriding (Regan, 1983, p. 286).

Those individuals that have inherent value have a basic right to be treated with respect. This right is equal for all individuals and it can "never" be overridden.

Inherent value should be distinguished from intrinsic value. Intrinsic value attaches to experiences, while inherent value attaches to beings that have experiences. In a discussion of moral agents, Regan clarifies what he means by "inherent value" in the following manner:

> The inherent value of individual moral agents is to be understood as being conceptually distinct from the intrinsic value that attaches to the experiences they have (for example, their pleasures or preference satisfactions), as not being reducible to values of this latter kind, and as being incommensurate with these values (Regan, 1983, p. 235).

We see here that Regan criticizes both hedonistic utilitarianism and preference utilitarianism. Both kinds of utilitarianism focus on the intrinsic value of particular experiences. Hedonistic utilitarianism stresses the intrinsic value of pleasures, while preference utilitarianism emphasizes the intrinsic value of preference-satisfaction.

Regan criticizes hedonistic utilitarianism with a cup metaphor, illustrating that the important thing is inherent value, not the quantity of pleasure or pain:

> One way of diagnosing its fundamental weakness is to note that it assumes that both moral agents and patients are, to use Singer's helpful terminology, *mere receptacles* of what has positive value (pleasure) or negative value (pain). They have no value of their own; what has value is what they contain (i.e., what they experience). An analogy might be helpful. Suppose we think of moral agents and patients as cups into which may be poured either sweet liquids (pleasures) or bitter brews (pains) (Regan, 1983, pp. 205–206).

He continues:

> The postulate of inherent value offers an alternative. The cup (the individual) has a value *and* a kind that is not reducible to, and incommensurate with, what goes into the cup (e.g., pleasure) ... It's the cup, not just what goes into it, that is valuable (Regan, 1983, p. 236).

Individuals have inherent value, not their experiences.

But which individuals have inherent value? In the quotation above (from Regan, 1983, p. 206), we saw that Regan talks about "moral agents" and "moral patients." Moral agents are individuals that can be held morally accountable. They have many sophisticated abilities, in particular the ability to apply moral principles and to freely choose or fail to choose to act according to these principles (Regan, 1983, p. 151). Moral patients (in the sense central to Regan's book) are individuals that are conscious, sentient, and have other cognitive and volitional capabilities, but do not have the moral ability characterizing moral agents. They cannot be held morally accountable. Examples of moral patients are "human infants, young children, mentally deranged or enfeebled of all ages" but also many animals (Regan, 1983, p. 153). With regard to inherent value, no difference exists between moral agents and moral patients. Regan maintains:

> *The validity of the claim to respectful treatment, and thus the case for recognition of the right to such treatment, cannot be any stronger or weaker in the case of moral patients than it is in the case of moral agents.* Both have inherent value, and both have it equally; thus, both are owed respectful treatment, as a matter of justice (Regan, 1983, p. 279).

Both moral agents and moral patients have inherent value, and both moral agents and moral patients have it equally. This means that no grading in inherent value exists between human beings and animals. It is not the case that human beings have a higher inherent value and animals a lower inherent value. Human beings and animals have equal inherent value. On the basis of this equal inherent value, moral agents and moral patients—including animals—have a right to respectful treatment. Regan stresses that this view that they have an equal right to respectful treatment is a matter of justice. This shows that Regan views his theory of rights as a theory of justice rather than a theory of moral obligation in general.

C. Subjects-of-a-Life

We have seen that all individuals, whether moral agents or moral patients, have equal inherent value. Many animals are moral patients, but which animals more precisely? Regan's answer is: those animals that are subjects-of-a-life.

> Individuals are subjects-of-a-life if they have beliefs and desires; perception, memory, and a sense of the future, including their own future; an emotional life together with feelings of pleasure and pain; preference- and welfare-interests; the ability to initiate action in pursuit of their desires and goals; a psychophysical identity over time; and an individual welfare in the sense that their experimental life fares well or ill for them, logically independently of their utility for others and logically independently of their being the object of anyone else's interests (Regan, 1983, p. 243).

It is not clear which animal species Regan refers to, but he indicates that at least all mentally normal mammals one year old or more fulfill the criteria (Regan, 1983, p. 247). This means that Regan refers to a more limited set of animals than Singer. We saw above that in Singer's view all sentient animals have interests that should be considered equally. Some sentient beings do not fulfill Regan's criteria of a subject-of-a-life. On the other hand, Singer's concept of a person comes quite close to Regan's concept of a subject-of-a-life. According to Singer, a person is an individual that is rational and self-conscious. These two abilities are not included in Regan's list, at least not explicitly. They seem to represent somewhat higher requirements. This implies that Singer's concept of a person covers a more limited set of animals than Regan's concept of a subject-of-a-life. Some animals that are subjects-of-a-life in Regan's sense are not persons in Singer's sense.

These differences between Singer and Regan appear to have implications for their level of protection. Sentient animals that are not subjects-of-a-life seem to have less protection in Regan's system than in Singer's. They do not have any rights according to Regan (although this is not logically precluded, as Regan points out (Regan, 1983, p. 246)), while their interests should be considered equally according to Singer. On the other hand, animals that are subjects-of-a-life in Regan's system seem to have stronger protection than those who are sentient beings but not persons in Singer's system, since rights seem to be less easily overridden than interests in cases of conflict.

D. The Rights of Animals

We saw above that Regan primarily views his theory of animal rights as a theory of justice. He can be interpreted as defending interspecies justice. Let us investigate what rights animals have.

I have already mentioned the basic right to respectful treatment. In a way, Regan can be said to apply Kant's categorical imperative to animals (Kant did not do this himself; as we have seen above, he had quite a different view of animal ethics). As human beings should not be treated as means only but always as ends-in-themselves, so should animals that are subjects-of-a-life. They should not be instrumentalized, but should be treated with respect (*cf.* Regan, 1983, p. 249). They are not to be viewed or treated as "mere receptacles." Those who have inherent value should be treated in ways that respect their inherent value. Regan calls this "The Respect principle" and this principle can "never" be overridden (Regan, 1983, pp. 248, 286). This means that it is an absolute principle and that the animal right to be treated with respect is an absolute right.

Another important principle is "the harm principle." This principle is less fundamental than the respect principle, because it can be derived from it (Regan, 1983, p. 262). Animals that are subjects-of-a-life have a right not to be harmed. "Harm" can mean two different things, however. First, harm can be inflictions that diminish quality of life and overall welfare. Inducing pain in an animal is a clear example. Second, it can mean deprivations. The animals are denied opportunities for doing what will lead to satisfaction. For example, death deprives an individual of the opportunities of life (Regan, 1983, pp. 94–99).

In contrast to the respect principle, the harm principle can be overridden. The right not to be harmed is only a *prima facie* right (Regan, 1983, p. 287). A presupposition for this *prima facie* character of the harm principle is that harms can be compared, and some harms are "comparable" while others are not. By "comparable harms" Regan means harms that detract equally. One option is that one and the same individual is subjected to comparable harms at different times, another that different individuals are subjected to comparable harms (Regan, 1983, pp. 303–304). This comparability opens up the possibility of balancing harms among individuals. The right of one individual not to be harmed may override the right of another. Regan suggests certain principles for handling such conflicts between rights holders. Let us turn to an analysis of these principles.

E. Conflicts between Rights Holders

I start by mentioning a principle that Regan explicitly rejects, namely the "minimize harm principle." This principle means that we should minimize the total aggregate of harm (of innocent individuals). Regan characterizes it as a consequentialist principle and objects to it in a way similar to the way he objects to hedonistic utilitarianism, although in this case it is a matter of aggregating harms and benefits instead of pleasures and pains. According to Regan, the minimize harm principle is wrong, because like hedonistic utilitarianism it treats moral agents and moral patients as mere receptacles instead of beings with inherent value (Regan, 1983, pp. 301–303).

Regan suggests two other principles that should govern our handling of conflicts of rights holders. First, he suggests a "minimize overriding principle" ("miniride principle"). This means that when the few and the many are harmed in a *prima facie* comparable way, then we should override the rights of the few instead of the rights of the many (Regan, 1983, pp. 305–307).

Second, Regan proposes a "worse-off principle," which is applicable when individuals are harmed in a *prima facie* non-comparable way, for example when the harm of the few would make them worse-off than any of the many. In this case, the rights of the many are overridden instead of the rights of the few that would be worse-off (Regan, 1983, pp. 307–312).

Let us take a closer look at the worse-off principle as applied in situations involving both animals and human beings. Regan illustrates this with an analogy of a lifeboat:

> Imagine five survivors are on a lifeboat. Because of limits of size, the boat can only support four. All weigh approximately the same and would take up approximately the same amount of space. Four of the five are normal adult human beings. The fifth is a dog. One must be thrown overboard or else all will perish. Whom should it be? (Regan, 1983, p. 285).

Later he gives the following answer:

> The rights view's answer is: the dog. The magnitude of the harm that death is ... is a function of the number and variety of opportunities for satisfaction it forecloses for a given individual, and it is not speciesist to claim that the death of any of these humans would be a prima facie greater harm in their case than the harm death would be in the case of the dog. Indeed, numbers make no difference in this case. A million dogs ought to be cast overboard if that is necessary to save the four normal humans (Regan, 1983, p. 351).

Elsewhere he clarifies:

> Death for the dog, in short, though a harm, is not comparable to the harm that death would be for any of the humans... Our belief that it is the dog that should be killed is justified by appeal to the worse-off principle (Regan, 1983, p. 324).

This means that Regan accepts that human harm may override animal harm in lifeboat situations because of the number and variety of opportunities for satisfaction that it forecloses. The human beings would be worse-off. In order to make his point, Regan stresses that even a million of dogs should be cast overboard in order to save the human beings.

This is a critical aspect of Regan's rights view. How impartial is he? Is it truly "not speciesist" to maintain that the right of four human beings not to be harmed overrides the right of millions of dogs not to be harmed? The premise is that the number and variety of opportunities for satisfaction is ethically crucial. We might ask, however, why such a premise is not speciesist. It appears to be a very "human" point of view.

Regan obviously does not view animal experiments, for example medical experiments, as lifeboat situations. The reason is that laboratory animals and human patients do not suffer comparable harms, because the animals would not be harmed were they not used in the experiments. In lifeboat situations, all involved animals and human beings would be harmed, that is, die, if nobody were cast overboard. In Regan's view, the minimize harm principle is unacceptable. According to that principle, it could be right to cause at least minor animal harm to prevent major human harm such as life-threatening human diseases.

F. Using Animals in Research

Animals that are subjects-of-a-life should not be instrumentalized, but should be treated with respect. In Regan's opinion, this excludes, for example, animal agriculture, hunting and trapping, and animal experimentation. But he is not against all uses of animals. For example, he does not seem to exclude companion animals, if they are treated well (Regan, 1983, pp. 330–398).

The focus here is on animal experimentation. Regan exemplifies harm in research by referring to burns, poisoning, surgery, and sensory deprivation. He stresses that it is not sufficient to provide anesthesia and post-operative pain relief. The reason is that it is not the pain or suffering that is important but the harm in terms of reduced welfare opportunities and untimely death (Regan, 1983, pp. 387–388). Regan points out, however, that "the rights view is not against research on animals, if this research does not harm these animals or put them at risk of harm" (Regan, 1983, p. 387). It is against harmful research only. Field studies, for example, could be acceptable. In order to find treatments for human diseases non-animal alternatives should be sought (Regan, 1983, p. 388).

The animals we talk about as moral patients with rights based on inherent value are those that are subjects-of-a-life. We saw above that these include at least all mentally normal mammals one year old or more. What about using mammal embryos or fetuses? Here Regan maintains that these can be used provided that such use would not foster attitudes that would sanction disrespectful treatment of animals that have rights. This can only be expected when scientists stop using mammals in their later stages of life (Regan, 1983, pp. 390–392).

G. Conclusion

In sum, Regan's animal rights prototype is opposed to all animal experimentation. To use animals that are subjects-of-a-life in experiments would be unjust. It would violate their rights.

4. Strong Human Priority

Carl Cohen defends animal experimentation against the criticisms of Regan and Singer. This was first done in an influential article from 1986 in *New England Journal of Medicine* entitled "The case for the use of animals in biomedical research." This article will be the main source in the following analysis (although I will refer to the reprint in Cohen 1994). An additional source consists of the parts written by Cohen in the book *The Animal Rights Debate* published together with Regan in 2001.

A. The Benefits of Animal Experimentation

Cohen has a very positive view of the benefits of animal experimentation:

> Every disease eliminated, every vaccine developed, every method of pain relief devised, every surgical procedure invented, every prosthetic device implanted—indeed virtually every modern medical therapy is due, in part or in whole, to experimentation using animals (Cohen, 1994, p. 261).

He concludes:

> If the morally relevant differences between humans and animals are borne in mind, and if all relevant considerations are weighed, the calculation of long-term consequences must give overwhelming support for biomedical research using animals (Cohen, 1994, p. 261).

This statement about the consequences of animal experimentation is a direct criticism of Singer. Cohen accuses Singer of underestimating the benefits of animal experimentation and ending up in an inaccurate consequentialist balancing of costs and benefits. Cohen here argues in term of consequences, despite the fact that he elsewhere in the article argues in terms of rights. This indicates that he embraces a kind of mixed ethical theory. He also stresses the existence of morally relevant differences between human beings and animals. Let us investigate both these aspects in more detail.

B. Criticism of Animal Rights

Cohen's argument for morally relevant differences is found in his criticism of Regan's animal rights view. He finds the notion of animal rights conceptually flawed. The concept of rights cannot be applied to animals. He argues:

> This much is clear about rights in general: they are in every case claims, or potential claims, within a community of moral agents. Rights arise, and can be intelligibly defended, only among beings who actually do, or can, make moral claims against another .... Animals (that is, nonhuman animals, the ordinary sense of that word) lack this capacity for free moral judgment. They are not beings of a kind capable of exercising or responding to moral claims. Animals therefore have no rights, and they can have none. The holders of rights must have the capacity to comprehend rules of duty, governing all including themselves... In conducting

research on animal subjects, therefore, we do not violate their rights, because they have none to violate (Cohen, 1994, p. 254).

Animals lack moral ability. They are not members of a community of moral agents. Thus, animals can have no rights, because rights can only be attributed to members of a moral community. Human beings belong to such a community. They have moral obligations and, correspondingly, rights. Animals have no obligations, and therefore no rights. Rights arise only among beings that make moral claims against each other.

According to Cohen, human beings have rights, never animals. But what is the foundation of rights? He mentions several different possibilities such as divine gift, human moral community, direct intuitive recognition, and natural evolutionary development (Cohen in Cohen and Regan, 2001, pp. 32–33; see also Cohen, 1994, p. 254). Cohen does not take a definite stand on this point, but stresses the human locus of rights (Cohen in Cohen and Regan, 2001, p. 34). Human beings are self-legislative and members of moral communities (Cohen, 1994, p. 255).

C. Obligations to Animals

That animals have no rights does not mean that we have no obligations to animals. Cohen states:

> It does not follow from this, however, that we are morally free to do anything we please to animals. Certainly not. In our dealings with animals, as in our dealings with other human beings, we have obligations that do not arise from claims against us based on rights. Rights entail obligations, but many of the things one ought to do are in no way tied to another's entitlement ... In our dealings with animals, few will deny that we are at least obliged to act humanely—that is to treat them with the decency and concern that we owe, as sensitive human beings, to other sentient creatures (Cohen, 1994, p. 255).

We have moral obligations to animals as sentient creatures, even though they have no rights. The reason is that they can be harmed, suffer, and feel pain. Cohen clarifies:

> The obligation to act humanely *we owe to them* even though the concept of a right cannot possibly apply to them (Cohen in Cohen and Regan, 2001, p. 29).

In Cohen's view, animals have moral status, but this status is lower than the moral status of rights bearers, that is, human beings. We see here a difference compared to Carruthers. According to Carruthers, we have only indirect duties to animals. According to Cohen, we have direct duties to them. I interpret these duties as being based on consequentialist considerations. Our acts may make them suffer.

Thus, Cohen embraces a theory that is a mix of a rights-based approach and consequentialism. Rights emerge in a human moral community, but morality comprises more than rights. We may also have obligations beyond those that are linked to rights. Cohen gives several examples of obligations that do not correspond to rights. Such obligations may arise from internal commitments (for example, teachers' obligations to students), from differences of status (for example, adults' obligations when playing with young children), from special relationships (for example, a father's obligation to pay college tuition for his son), and from particular acts or circumstances (for example, obligations due to a special kindness done to oneself) (Cohen, 1994, pp. 255–256). It is not clear from Cohen's article how these obligations are to be justified, whether deontologically or consequentially. However, it appears that our obligations to animals are to be justified in a consequentialist manner, since he stresses that we have these obligations because animals can suffer. In addition, his willingness to balance human benefit and animal suffering indicates that his consequentialism concerns not only animals but also human beings. I draw the conclusion that in the case of human beings Cohen has a mixed approach, a rights-based one but also a consequentialist one.

D. Speciesism as Crucial for Right Conduct

Cohen criticizes strongly Singer's view of speciesism. He confesses boldly that: "I am a speciesist. Speciesism is not merely plausible; it is essential for right conduct" (Cohen, 1994, p. 259). Cohen does not define "speciesism," but he would probably accept Singer's definition in terms of giving "greater weight to the interests of members of their own species when there is a clash between their interests and the interests of those of other species" (Singer, 1993a, p. 58). Cohen criticizes the analogy between speciesism and racism suggested by Singer. Racism is wrong because human beings really are equal. No morally relevant differences among human ethnic groups can be found, but there exist morally relevant differences between human beings and animals. Animal pains are to be weighed but animal pains and human pains are not to be weighed equally. We have special obligations to human beings that we do not have to animals (Cohen in Cohen and Regan, 2001, pp. 62–63).

A possible objection to Cohen—to which he tries to respond—is that newborns and severely mentally retarded individuals seem to be excluded from the community of rights holders, because they do not have moral ability in the rational sense that they can respect the rights of others. In Cohen's opinion, this is no real problem, since these individuals are members of the moral community, even if they do not have a moral ability in the full sense. Moral ability is not a test to be applied to human beings one by one. The issue is one of "kind." The human species is a kind that has this moral ability.

E. Views on Replacement and Reduction

I started my presentation of Cohen by pinpointing his positive assessment of the human benefits of animal experimentation. Let me finish by clarifying Cohen's views on the possibility of using non-animal alternatives in biomedical research and on reducing the number of animals used.

According to Cohen, no other methods exist that can replace testing a drug or a procedure on living organisms. Sooner or later the drug or procedure has to be tested on a whole living being. If we did not carry out the testing on animals, valuable research would be blocked or we would have to experiment directly on human beings. Both options are closed if we take our obligations to our fellow human beings seriously (Cohen, 1994, pp. 261–262).

Cohen is also critical of reduction. If we reduce the number of animals, the advancement of medicine would slow down and that would be contrary to our obligations to the other members of our moral community. Cohen instead argues for increasing the number of animals in biomedical research (Cohen, 1994, pp. 262–263).

F. Conclusion

In sum, Cohen can be viewed as a proponent of a "strong human priority" prototype. According to this prototype, human research interests always have higher priority than animal interests, given that we take animal welfare into consideration. We should try to minimize animal suffering as long as it does not seriously hinder research.

5. Weak Human Priority

A prominent proponent of the weak human priority view is Mary Midgley. In this presentation I will focus on her book *Animals and Why They Matter* from 1983. This book is on animal ethics in general. It has no special focus on animal experimentation, but it has some references to animal experimentation, and it is quite easy to see the implications of her view of animal ethics for this issue. Her position represents a middle course in the debate between Carruthers and Cohen on the one hand, and Singer and Regan on the other.

A. Some Animal Experiments Are More Justified than Others

Midgley's view on animal experimentation is described as follows:

> It must emerge that some experiments are much more justifiable than others. The feat of justification cannot, in any case, be performed merely by raising an umbrella marked "Science." It demands attention to the actual benefits which can reasonably be expected, and a serious comparison of the conflicting values involved (Midgley, 1983, p. 28).

Midgley is against accepting all animal experiments but she is also against not accepting any animal experiments. Instead animal experiments must be assessed on a case-by-case basis, weighing the pros and cons. Some animal experiments are more justified than others, but it is not clear to what extent Midgley finds animal experiments ethically acceptable. Before we discuss different possible interpretations, we must investigate Midgley's view of speciesism.

B. Weak Normative Speciesism

Midgley appears to understand speciesism as "preference for one's own species" (Midgley, 1983, p. 104) or as "relative disregard of other creatures" (Midgley, 1983, p. 106). This is compatible with Singer's definition mentioned above. However, like Cohen she criticizes Singer's view that speciesism is a kind of prejudice on a par with racism and sexism (Midgley, 1983, pp. 96–97). Instead she argues that we should take speciesism seriously in normative ethics, although she renounces complete dismissal of animal interests.

Midgley neglects to make an explicit distinction between descriptive speciesism and normative speciesism. Descriptive speciesism is a statement about human beings favoring their species, while normative speciesism is an ethical view that human beings should or are allowed to favor their species. "Favoring" means prioritizing our species when there is a conflict either in terms of human and animal interests or in terms of obligations to human beings and animals. Midgley urges us to take descriptive speciesism seriously in normative ethics. She uses descriptive speciesism in an argument for normative speciesism.

Midgley does not accept general, complete, or nearly complete normative speciesism. In distinction to such "strong" normative speciesism—defended by Cohen—she instead embraces what I would call "weak" normative speciesism, that is, the view that we should or are allowed to favor our species, but only to some extent.

> There does seem to be a deep emotional tendency, in us as in other creatures, to attend first to those around us who are like those who brought us up, and to take much less notice of others. And this, rather than some abstract judgement of value, does seem to be the main root of that relative disregard of other creatures which has been called "speciesism." I shall suggest in a moment that this natural tendency, though real, is nothing like so strong, simple and exclusive as is sometimes supposed, and has neither the force nor the authority to justify absolute dismissal of other species (Midgley, 1983, p. 106).

Before we investigate more precisely how weak Midgley's normative speciesism is, let us look at the kind of reasoning supporting this view.

C. Arguments for Weak Normative Speciesism

Midgley presents two arguments in favor of weak normative speciesism. The first is that it is natural. She argues:

> An emotional, rather than rational, preference for our own species is ... a necessary part of our social nature, in the same way that a preference for our own children is, and needs no more justification .... The natural preference for one's own species [is not] a product of culture. It is found in all human cultures, and in cases of real competition it tends to operate very strongly .... The natural preference for one's own species does exist. It is not, like race-prejudice, a product of culture. It is found in all human cultures, and in cases of real competition it tends to operate very strongly. We can still ask, however, how far it takes us (Midgley, 1983, p. 104).

We are creatures characterized by social bonding. We have an innate tendency to care more for our family than strangers, and more for human beings than non-human animals. She suggests that this psychological propensity to social bonding is a result of biological evolution. It is a preference like the preference for our children and it needs no more justification. This is obviously an argument from descriptive speciesism to normative speciesism by means of the concept of naturalness. However, at the end of the quotation, Midgley indicates that reference to this propensity is of only limited use in ethics. This indicates that her normative speciesism is only weak. Descriptive speciesism cannot be used to justify strong normative speciesism (see below).

Midgley's second argument for weak normative speciesism is that it is crucial for human happiness. She stresses that our family and species preferences are

> an absolutely central element in human happiness, and it seems unlikely that we could live at all without them. They are the root from which charity grows. Morality shows a constant tension between measures to protect the sacredness of these special claims and counter-measures to secure justice and widen sympathy for outsiders. To handle this tension by working out particular priorities is our normal moral business. In handling species conflicts, the notion of simply rejecting all discrimination as *speciesist* looks like a seductively simple guide, an all-purpose formula. I am suggesting that it is not simple, and that we must resist the seduction (Midgley, 1983, p. 103).

We would not be happy, if we ignored social bonding and tried to be entirely impartial. "We are bond-forming creatures, not abstract intellects" (Midgley, 1983, p. 102). Moreover, we would not be more virtuous if we ignored social bonding in the way Singer suggests. Parents who do not pay special attention to their children compared to other children are bad parents and their children will not develop into caring individuals. Social bonding is the basis for developing the virtue of charity.

D. The Precise Meaning of Midgley's Weak Normative Speciesism

In the literature, we find different interpretations of the extent of Midgley's speciesism. David DeGrazia and Rosalind Hursthouse interpret her differently, but both misinterpret her, at least partly. DeGrazia interprets Midgley as if she sometimes argues in favor of a "general discounting of animals' interests" and sometimes not (DeGrazia, 1996, p. 64). I think it is wrong ever to interpret her as defending a general discounting. I can find no such statements in her book. On the contrary, we saw above that Midgley does not view the natural speciesist tendency to be so strong that it justifies "absolute dismissal" of animals (Midgley, 1983, p. 106). Hursthouse, on the other hand, is partly correct when she interprets Midgley as arguing that "speciesism is sometimes justifiable, sometimes not" (Hursthouse, 2000, p. 132). However, she neglects to note that in Midgley we find a tendency toward "commonly" (my analytic term) rather than merely "sometimes." Otherwise Midgley's statement against "absolute dismissal" would be strange; it presupposes that a quite substantial dismissal is acceptable. The whole point of stressing the ethical obligations arising from social bonding is that we are commonly allowed to use animals for the benefit of human beings. Midgley is not quite clear in this regard. The term "commonly" may be used to cover a range from "in a majority of cases" to "almost always." In order to make real sense, the statement "speciesism is commonly ethically justifiable" must be related to the actual quantity of animal use in the world. Moreover, from an ethical point of view there may be qualitatively different uses of animals. Some uses of animals may be more ethically acceptable than others. For example, some animal experiments with expected high medical benefit for human beings might be more justified than very intensive farm animal production with low animal welfare. But it is almost impossible to predict what the outcome would be if case-by-case assessments of all animal uses were carried out according to this view. The statement therefore lacks precise meaning, but it does give a hint. It suggests a direction for our imagination.

When I use the term "commonly" in relation to Midgley's views in this chapter and coming chapters, this lack of precision should be kept in mind.

E. No Simple Formula for Priority-Setting

Midgley's case-by-case approach implies a criticism of the "concentric circle view" that fixed priorities exist between our duties. Michael P. T. Leahy, for example, appears to think in terms of concentric circles, arguing that we always have stronger moral obligations to human beings than to animals (Leahy, 1991, p. 172). Midgley states on the other hand (without referring to Leahy):

We might try, for instance, a series of concentric circles .... But at once we see that the order of the circles is not at all certain. At each point we may want to reverse it, or be dissatisfied with either order. Further groupings constantly occur to us, and, at every stage, it seems that some groupings are more important for some purposes, some for others. The concentric arrangement will not work at all. We must imagine instead a set of overlapping figures of varying shapes, representing various *kinds* of claims and loyalties ... There is obviously no simple formula for determining priority among these distinct kinds of claims, and moral philosophies like Utilitarianism which try to make the job look simple can only deceive us. Each culture, and each individual, must and does work out a map, a quite complex set of principles for relating them (Midgley, 1983, pp. 28–30).

The key point here is that there is "no simple formula for determining priority among these distinct kinds of claims." Midgley links this statement to an explicit criticism of "utilitarianism" but also of other "moral philosophies" that give us simple solutions to complex ethical problems. No simple formula or algorithm exists that provides definite answers. This indicates a casuistic tendency in Midgley, although her view is not purely casuistic. She also talks about "principles" for relating various complex moral claims, but it remains unclear what these "principles" are.

F. Relational Ethics *and* Interspecies Justice

Singer stresses sentience as a property that confers moral standing. Midgley also accepts this (Midgley, 1983, p. 96). In addition, she accepts relational properties as ethically relevant. Due to social bonding we stand in special relations to our family, other human beings, and even animals. These special relations give rise to special obligations (Midgley, 1983, pp. 98–111).

As I interpret Midgley, we have commonly stronger moral obligations to our children than to strangers and to human beings compared to animals. Sometimes, however, obligations to strangers may outweigh obligations to our children, and sometimes obligations to animals may outweigh obligations to human beings. Moreover, we commonly have stronger moral obligations to animals that are in our care—for example, pets and farm animals—than to wild animals, but sometimes obligations to wild animals may outweigh obligations to animals in our care.

We may even form "mixed communities" of human beings and animals (Midgley, 1983, p. 112). Midgley points out:

It is one of the special powers and graces of our species ... to draw in, domesticate and live with a great variety of other creatures. No other animal does so on anything like so large a scale. Perhaps we should take this peculiar human talent more seriously and try to understand its workings (Midgley, 1983, p. 111).

It appears that Midgely proposes a kind of relational ethics. Such an ethics is characterized by the recognition of relational properties as morally relevant. Similar attention to relational properties can be seen in, for example, the writings of Nel Noddings and Mary Ann Warren (Noddings, 1984; Warren, 1997, pp. 122–177, 224–240). In Midgley we do not find a pure version of relational ethics, that is, a view according to which only relational properties are morally relevant. We have special obligations to those we stand in special relations to, but we should also take impartial justice—for example, interspecies justice—into account. Midgley states:

> Morality shows a constant tension between measures to protect the sacredness of these special claims and counter-measures to secure justice and widen sympathy for outsiders. To handle this tension by working out particular priorities is our normal moral business (Midgley, 1983, p. 103).

We are to take measures to protect special relations but also to take countermeasures against unjustly favoring our species or kin. Working out particular priorities is difficult but part of our "normal moral business."

G. Weak Human Priority in Animal Experimentation

Applied to animal experimentation, Midgley's weak normative speciesism implies that, in cases of conflict, human research interests—under the presumption that they will be of human benefit—commonly, but not always, outweigh animal interests. This makes Midgley a proponent of a "weak human priority" prototype in the ethics of animal experimentation. This prototype differs from Cohen's strong human priority prototype according to which human research interests always outweigh animal suffering, but also from Singer's view that human research interests almost never outweigh animal suffering. However, as in the description of weak normative speciesism, the term "commonly" lacks precise meaning and indicates only a direction for our imagination. Central to Midgley's view is that animal experiments should be assessed on a case-by-case basis.

H. Conclusion

According to the strong human priority prototype, animal interests may influence *how* research is carried out but not *whether* it should be carried out. According to the "weak human priority" prototype, on the other hand, animal interests may influence *how* research is carried out and *whether* it should be carried out. Human research interests commonly—whatever that would mean more precisely—have higher priority than animal interests, but not always.

Animal experiments have to be justified on a case-by-case basis, taking animal interests seriously.

6. A Spectrum of Views

Let me finally summarize the prototypes by starting with the most positive view and finishing with the most negative one.

The human dominion prototype implies that all animal experimentation with expected human benefit is acceptable, provided that animal welfare is taken seriously for the sake of animal lovers. The strong human priority prototype accepts all animal experimentation with expected human benefit, provided that animal suffering is minimized for the sake of the animals themselves. Animal interests have no impact on whether animal experiments are carried out, only on how. In the weak human priority prototype, animal interests may influence both whether and how animal experiments are to be carried out. According to the prototype of equal consideration of interests, most animal experiments are unacceptable. Expected human benefit almost never outweighs animal harm. The animal rights prototype accepts no animal experimentation whatsoever.

Three

THE CASE FOR "WEAK HUMAN PRIORITY"

Which prototype is the most well-founded? This is the key question of this chapter. I will start this discussion by giving a few examples of how legal regulation regarding animal experimentation can be categorized in relation to the five prototypes. This is important since it puts the philosophical discussion of these prototypes into a broader societal perspective. Conversely, it also puts legal regulation into a broader philosophical perspective.

Ethics and law are two different things, but they can be related in different ways. Ethical considerations may be used to support legal regulation, but they may also be used to criticize such regulation. When relating the five ethical prototypes of animal experimentation to law, I just want to point out some similarities. Legal regulations in different countries may come more or less close to particular ethical prototypes.

1. Legal Regulation and the Five Prototypes

As mentioned in the Introduction, regulations generally—with the exception of the Animal Welfare Act of the United States—apply to research involving vertebrates. In the United States, the Animal Welfare Act does not include mice, rats, and domestic birds (Animal Welfare Act, 1985, Section 2 (g)), while the regulations of the Public Health Service include these species and all other vertebrates (Public Health Service, 1986, Section 3A). Since the latter regulations apply only to federally funded research, no regulation exists for privately funded research on mice, rats, and birds.

In general, the national policies in many countries encourage researchers to consider alternatives to the use of animals and to minimize the use of animals when such use is required. They also stress the importance of minimizing animal suffering in experiments and in living conditions. However, differences in institutional structure can be found among different countries. In, for example, the United States, Canada and my own country, Sweden, the policies are enforced by local institutional committees; in Germany by regional agencies; and in the United Kingdom and France by a national agency.

In the United States and Canada, the regulations say nothing about weighing but express instead the strong human priority position. Animal interests count, but it is not suggested that they may outweigh human research interests. The Animal Welfare Act of the United States (1985, Section 1) maintains that the use of animals is instrumental in certain research and education for advancing knowledge of cures and treatment for diseases and injuries which afflict both humans and animals.

But the Act also states that an aim is

> to insure that animals intended for use in research facilities or for exhibition purposes or for use as pets are provided humane care and treatment.

Similarly, the Public Health Service Policy on Humane Care and Use of Laboratory Animals states:

> Proper use of animals, including the avoidance or minimization of discomfort, distress, and pain when consistent with sound scientific practices, is imperative. Unless the contrary is established, investigators should consider that procedures that cause pain or distress in human beings may cause pain or distress in other animals (Public Health Service, 2002, Section IV).

By contrast, the regulations in the United Kingdom and Sweden embody a weak human priority position, which endorses a general balancing principle of forbidding research when human benefit is not sufficient to justify animal suffering. In the United Kingdom, the Secretary of State, through a set of inspectors and independent assessors, is required to

> weigh the likely adverse effects on the animals concerned against the benefit likely to accrue as the result of the programme to be specified in the licence (Animals (Scientific Procedures) Act, 1986, section 5.4).

In Sweden, the Animal Welfare Ordinance states that the regional ethics committees on animal experimentation in assessing particular research projects should proceed as follows:

> When considering specific cases the committee shall weigh the importance of the experiment against the suffering inflicted on the animal (Animal Welfare Ordinance 1988 (with later revisions), section 49.1).

Some of the other European Union countries follow more closely the Council Directive 86/609/EEC "on the approximation of laws, regulations, and administrative provisions of the Member States regarding the protection of animals used for experimental and other scientific purposes." It states:

> Where it is planned to subject an animal to an experiment in which it will, or may, experience severe pain which is likely to be prolonged, that experiment must be specifically declared and justified to, or specifically

authorized by, the authority. The authority shall take appropriate judicial or administrative action if it is not satisfied that the experiment is of sufficient importance for meeting the essential needs of man or animal (Article 12, section 2).

This implies a somewhat less weak human priority position, indicating that the balancing principle holds only in cases of severe animal suffering (*cf.* Brody, 1998, p. 24).

In conclusion, we see that in some important research countries, the regulation ranges from a strong human priority position to a weak human priority view. However, the human dominion prototype, the prototype of equal consideration of interests, and the animal rights prototype, are not represented.

This result is interesting. In all these countries, legal regulation comes close to some kind of middle course in the ethics of animal experimentation. This result should be compared to the result of the MORI poll in the United Kingdom described in Chapter One, which showed that the general public, at least in that country, tends toward a middle course. These results do not mean that a middle course is necessarily right or even the most well-considered position. However, the results of the legal analysis and the poll justify a deeper analysis of the possible arguments for such a middle course and of the presumptions involved. The general academic debate on the ethics of animal experimentation has been dominated by extreme views, in particular by those that are very critical, such as Singer's view. In legal acts, arguments are seldom put forward for particular regulations, although we can find such arguments in governmental reports written in preparation for proposals of legislation. To analyze such reports is far beyond the purpose of this book, but a need definitely exists for investigating the arguments and presumptions of middle positions in more detail.

2. Ethical Theory

I will now discuss a few central aspects of the five ethical prototypes. What is at stake is nothing less than several classical issues in moral philosophy. I focus on ethical theory and key metaphors, but also on the implicit presumptions of these prototypes regarding the following philosophical issues: intrinsic versus relational properties, reason versus feelings, impartiality versus special obligations, and the relation of "is" and "ought" (there will be some repetition but this is necessary for the discussion).

Let us start with ethical theory. Your fundamental choice of ethical theory may influence your view of animal experimentation. If you adopt one general approach to ethics instead of another, this may have implications for your view of the ethics of animal experimentation. But it can also be the other way around. If you are attracted to a particular type of view on animal experimentation, you may appeal to different ethical theories in order to support it.

Before I discuss this, let me first briefly summarize the ethical frameworks of the five prototypes.

Carruthers has a contractualist approach, focusing on the reciprocal rights and duties of human beings in society. Animals have no moral status. They cannot take part in the moral contract. They have no direct rights, and human beings have no direct duties to them. Human beings have only indirect duties to animals due to respect for animal lovers and considerations of moral character (Carruthers, 1992).

Cohen has also a rights-based approach with regard to human beings, although it remains unclear what the foundations of rights are. He mixes this approach with a consequentialist approach regarding animals and human beings. He proposes a lexical order between direct obligations to human beings and direct obligations to animals. Non-trivial human research interests always outweigh animal interests (Cohen, 1994).

Midgley has a mixed ethical view, accepting both consequentialist and deontological considerations with regard to human beings and animals. She exhibits a casuistic tendency, stressing that decisions must be made on a case-by-case basis. The ordering between obligations to human beings and obligations to animals is not lexical but contextual. However, Midgley is not a pure casuist. She stresses the need to relate conflicting moral claims by means of a set of principles. In Midgley, we also find a tendency toward relational ethics, which is reminiscent of the ethics of care suggested by some feminist philosophers. She stresses that we may have special relations to our children, to our fellow human beings, and even to some animals. These relations confer special obligations. But Midgley is not proposing a pure relational ethics. Also in this regard she has a mixed view. We have special obligations due to special relations but these special obligations need to be balanced against justice. In the case of animal experimentation, our special obligations to human beings to develop medical treatments must be balanced against interspecies justice. In addition, Midgley appears to be influenced by Aristotle, supporting a virtue ethics with human happiness and flourishing as a goal. She does not mention *phronesis* (practical wisdom) explicitly, but this notion fits her approach very well (Midgley, 1983).

Singer is a preference utilitarian, stressing that in most cases human research interests do not outweigh animal interests. The expected benefits of most animal experiments do not outweigh the animal suffering they cause (Singer, 1993a; Singer, 1995).

Regan has a rights-based approach, although it differs from Carruthers's and Cohen's versions. They deny that animals have rights. Regan argues that animals that are subjects-of-a-life have rights. These rights are based on their inherent value. Regan recognizes some "lifeboat" situations in which conflicts arise between rights holders. In these situations, human beings may be prioritized. However, animal experiments are not a matter of lifeboat situations (Regan, 1983).

We see that if you embrace a rights-based approach, that view may be used in an argument that is positive toward animal experimentation (Carruthers and Cohen) or in one that is very critical (Regan). It depends on the more precise nature of the rights-based view.

If you adopt a utilitarian theory, you may also end up in different views on animal experimentation, depending on how you carry out the balancing of human interests and animal interests. If you have an optimistic view of the human benefits of animal experimentation, these benefits may be considered to outweigh animal suffering (Cohen). If you, on the other hand, have a pessimistic view of the human benefits, they may not compensate for the animal suffering (Singer).

Finally, if you have a mixed ethical view, this view may be used to support any standpoint regarding animal experimentation, positive or negative. It all depends on which types of views are mixed and on exactly what the mixture looks like. Cohen's main theory is rights-based, but he mixes it with a utilitarian balancing of human benefit and animal suffering. Midgley's "mixed ethics" is more complex. She combines consequentialist and deontological considerations. She exhibits a casuistic tendency but stresses also the importance of working out a set of principles for prioritizing different valid moral claims. She mixes relational ethics with ethical considerations of justice. We also find traces of virtue ethics.

Thus, your choice of ethical theory may determine your view of the ethics of animal experimentation.

If you are attracted to a particular type of view on animal experimentation, you may appeal to different ethical approaches in order to support that view. The prototypes only represent a few of many possible ways of doing so. You may use other ethical theories to support views coming close to the prototypes.

The human dominion prototype of Carruthers is contractualist. Non-prototypical versions may be based on other types of ethical theory. In the historical overview, we saw, for example, different theological attempts to justify human use of animals by reference to human beings as created in the image of God and thereby having a higher moral status than animals.

Cohen's strong human priority prototype is rights-based but includes also consequentialist considerations. Non-prototypical versions of this category of views may be based on pure consequentialist accounts.

The weak human priority prototype of Midgley is based on a mixed ethical theory. Non-prototypical versions may be based on many different types of ethical theories, consequentialist or deontological, but it is necessary that they allow a balancing of interests, rights, or obligations from one case to another.

Singer's prototype of equal consideration of interests is preference utilitarian. Non-prototypical versions may presuppose hedonistic utilitarianism. It is also possible to ground non-prototypical versions on non-utilitarian approaches. For example, we may assume a rights-based approach that assigns *prima facie* rights to both human beings and animals, and that balances these rights in cases of conflict.

Regan's animal rights prototype is rights-based with rights being based on the presumption of the inherent value of subjects-of-a-life. Non-prototypical versions may include also animals that are not subjects-of-a-life. They may even be biocentric, including not only animals but also plants.

3. Key Metaphors

Metaphors are vital in framing ethical issues. They determine how we think. A metaphor is a concept from one domain of experience (the source domain) that is used to structure our understanding of another domain (the target domain) (Lakoff and Johnson, 1980; Lakoff and Johnson, 1999; Nordgren, 2001, p. 17). For example, we may say: "Love is a journey." In this case, we use the concept of a "journey" to structure our understanding of "love." Other metaphors may structure our understanding of animal experimentation. Choosing one metaphor instead of another may have ethical implications. Let me pinpoint some metaphors related to the prototypes of animal experimentation.

The terms in which the five prototypes have been caught are all metaphors. Note that the designations are in three cases my suggestions only, not designations used by the proponents. This holds true for "human dominion," "strong human priority," and "weak human priority." Within the framework of each prototype, different metaphors are used by the proponents themselves.

A key issue in all five prototypes is whether animals have moral standing. "Standing" is a metaphor of position or, more precisely, upright posture. The metaphor suggests moral rank or status. A related metaphor is that of counting. "Counting" is a mathematical metaphor. Animals and animal interests are considered morally relevant. Animals have a moral claim upon us such that we have a duty to consider them in our ethical deliberations. The negative counterpart of counting is "discounting," which means not to be counted or to be counted less. From the perspective of moral imagination the choice is: should we or should we not make an extension of the moral standing metaphor and the moral counting metaphor from the prototypical case of human beings to the non-prototypical case of animals?

The term "dominion" in the designation "the human dominion prototype" is a metaphor of a kingdom. Human beings are viewed as rulers over animals, using them for different purposes. Of key importance in this prototype is to clarify differences in properties between rulers and ruled. Carruthers suggests that the relevant properties are consciousness and rationality.

The prototypes of strong and weak human priority exhibit other metaphors. To prioritize is "to place something or someone before something or someone else." It is a metaphor of ordering. "Strong" and "weak" are metaphors of physical strength.

The prototype of equal consideration of interests is based on a metaphor of a balance with equally heavy weights in each scale. The balancing metaphor is also central to Midgley's and Cohen's prototypes. In these prototypes human benefit in most cases (Midgley) or generally (Cohen) outweighs animal harm. All three prototypes also use an economic metaphor. What is to be balanced are benefits and costs. What are considered to be benefits and costs vary among the prototypes.

The animal rights prototype also uses an economic metaphor but of another kind. Rights are credits, while the corresponding obligations are debts. Cohen (1994)—and by implication, Carruthers (1992)—criticizes Regan's use of the term rights and argues that animals can have no rights, because rights can only be attributed to members of a moral community. Only human beings belong to such a community. Only they have moral obligations and, correspondingly, rights. Animals have no obligations, and therefore no rights. From the point of view of use of metaphor, we can note that while Regan (1983) extends the rights metaphor from the prototypical case of human beings to the non-prototypical case of animals, Cohen is not willing to do this. But Cohen is prepared to extend the obligation metaphor to include also obligations to animals. We have moral obligations to animals as sentient creatures, even though they have no rights. The reason is that they can feel pain and suffer.

If we accept that we have obligations to animals, the problem arises as to the ground of these obligations. Regan thinks that the ground is rights and that the ground of rights in turn is "inherent value." Cohen criticizes Regan for conflating two different meanings of the term: the inherent value that every human being has and the inherent value that every living being has. Cohen accepts the view that animals have inherent value in the latter sense, but rejects the idea that animals have an inherent value in the full human sense (Cohen and Regan, 2001, pp. 246–248). It is obvious that "inherent value" is also a metaphor or rather a mix of two metaphors. "Inherent" is a container metaphor, "value" an economic metaphor. The combined metaphor underlines that a creature—whether a human being or a non-human animal—is not purely instrumental to the interests of others. To use such a creature for one's own purpose is morally problematic and needs justification. With this in mind, the problem can be rephrased as a problem of how far the metaphor of inherent value should be extended. Should it be extended from the prototypical case of human beings to the non-prototypical case of animals? In that case, does the meaning of "inherent value" change when attributed to animals as compared to human beings? I agree with Cohen that the meaning changes.

I would also like to highlight Regan's metaphor of a cup and its content, which he uses in his discussion of Singer (Regan, 1983, pp. 206, 236). According to Regan, what has value is the subject-of-a-life, metaphorically described as a cup, while Singer and other utilitarians only recognize the value of experiences that the subject-of-a-life has, metaphorically described as the content of the cup, for example coffee.

Let me finally present the metaphors that govern the views of what ethics is all about. Basically, two options exist: the metaphor of an imaginary impartial observer and the metaphor of a partial participant. According to the former—which is the dominating one in philosophy—being ethical would imply taking an impartial stance. According to the latter, ethics is a matter of special obligations due to special relations in which we are involved. Singer, Regan, Carruthers, and Cohen view ethics as a matter of adopting an impartial stance. In Singer this is suggested by the idea of equal consideration of preferences or interests. Regan's idea that both human beings and animals have rights based on their inherent value is also in line with this. Carruthers and Cohen would argue that adopting the perspective of an impartial observer would have us recognize that the idea of rights is possible only within a human community. Midgley has a mixed view. To some extent, she adopts the metaphor of an imaginary impartial observer (interspecies justice), but tries to combine this with the metaphor of a partial participant. We have special relations to our children and other human beings, and to be ethical is to take this partiality seriously.

4. Intrinsic *and* Relational Properties

A key issue in all five prototypes concerns which natural properties are ethically relevant or—to put it differently—which properties confer moral standing or status.

Carruthers considers rationality and consciousness as ethically relevant, and concludes that animals cannot be subjects of direct moral concern since they do not possess these properties to the extent necessary for taking part in a moral contract. He even suggests tentatively the view that animal pain is nonconscious (Carruthers, 1992, pp. 189, 192–193).

Singer focuses on sentience—the ability to feel pleasure and pain—and on being a "person" with self-consciousness and rationality. He argues that the interests of all sentient beings should be considered equally, and that it is worse to treat a person badly than a non-person (Singer, 1993a; Singer, 1995).

Regan uses the concept of a subject-of-a-life and describes in detail the properties of such creatures. For example, they have beliefs, desires, perception, memory, a sense of the future, including their own future, an emotional life, an ability to initiate action in pursuit of their desires, and a psychophysical identity over time (Regan, 1983, p. 243). All subjects-of-a-life have inherent value and thereby rights.

Cohen focuses on moral ability and the existence of a moral community. Only human beings have this ability and are part of a moral community with reciprocal rights and duties. In addition, Cohen recognizes sentience as morally relevant. Animals that are sentient have some moral standing. He argues, however, that we have stronger moral obligations to members of the moral community—that is, human beings—than to sentient animals (Cohen, 1994).

In the presentation of non-prototypical versions of the human dominion view, we saw also other examples such as being created in the image of God or having linguistic ability.

All properties mentioned so far are intrinsic properties in the sense that they are properties that beings are believed to have in themselves, regardless of their relations to other beings.

Midgley also accepts sentience as an intrinsic property that confers moral standing (Midgley, 1983, p. 96). In addition to this intrinsic property, she accepts relational properties as ethically relevant. We are creatures characterized by a propensity to social bonding. We stand in special relations to our family, other human beings, and even to some animals—for example, pets and farm animals—and these special relations give rise to special obligations (Midgley, 1983, pp. 98–111).

Obviously empirical science cannot solve the problem of which properties are ethically relevant. Neither can it solve the religious problem of whether the belief that human beings in distinction to animals are created in the image of God is true. The former is an issue of ethical deliberation and the latter is a religious/metaphysical issue that is beyond the empirical domain.

However, empirical findings can undermine assumptions regarding the uniqueness of some properties by showing that the differences between human beings and animals are differences in degree rather than differences in kind. Findings in ethology indicate that non-human animals may to some extent be sentient, conscious, self-conscious, rational, and to some extent even have linguistic ability and moral ability (see Chapter Five). In addition, the theory of evolution explains on a general level why this can be expected. Human beings and other "higher" species have common origins.

These empirical findings together with the explanation of these findings given by the theory of evolution make it difficult to justify the discounting of animals with reference to specific intrinsic properties, although this might still be logically possible. We are not so different from animals with regard to intrinsic properties that we can discount them. I believe that in order to justify the discounting of animals we have to give feelings a higher status in ethical deliberation and also—as Baruch Brody points out (Brody, 2001, p. 142)—challenge the whole idea that we are, in general, morally committed to equal consideration of interests by accepting that we have special obligations to human beings compared to animals.

Both courses are taken by Midgley. She stresses that feelings together with reason should guide us ethically. She argues that not only intrinsic properties but also relational properties may confer moral status. Special relations give rise to special obligations. However, it is extremely important to note that Midgley is not defending a complete discounting of animals, only a limited one. Below I will discuss in more detail both the role of feelings in ethical deliberation and special obligations. I will also discuss the problem of relating ethics to natural properties, a problem that is often described in terms of bridging the is/ought gap (*cf.* Nordgren, 2002).

5. Reason *and* Feelings

A common view in the history of philosophy is that ethics should be a purely rational enterprise. Key proponents of this view are, for example, Descartes and Kant. Carruthers, Singer, Regan, and Cohen seem to stand in this tradition. Their arguments imply that reason can and should govern our feelings. Singer and Regan use rational argument to counteract speciesist feelings. Carruthers and Cohen uses reason to give a rational argument for speciesism.

Midgley, on the other hand, joins, for example, David Hume in acknowledging the relevance of feelings. Hume stated this view clearly: "Reason is, and ought only to be, the slave of the passions" (Hume, 1978, p. 415). Midgley would probably not go that far. Reason is equally crucial, but it has obvious limits. Reason cannot motivate us morally, but reason can determine how best to satisfy the desires. For example, desire may motivate me to go Stockholm to meet my family, but reason has to determine the best way to go there. However, Midgley does stress the importance of taking feelings seriously. Feeling and reason are complementary (*cf.* Midgley, 1981). As she points out: "We have to do justice to both feeling and thought. This means considering them together, and as aspects of the same process" (Midgley, 1983, p. 42). Midgley stresses in particular the feeling of social bonding and argues for taking relations due to social bonding seriously in normative animal ethics. She also urges us to take our feelings of sympathy toward animals seriously (1983, pp. 98–111).

"Feeling" can mean many different things, for example, emotion and desire. My aim here is not to analyze these concepts in detail, to clarify the differences between them, or to investigate their relation, but only to point out that feelings in a broad sense—according to Midgley—are important in ethical deliberation.

Midgley is not alone in stressing the importance of feeling in reasoning. Let me give two examples. The cognitive neurologist Antonio Damasio stresses in his book *Descartes' Error: Emotion, Reason, and the Human Brain* (1994) the importance of emotion in reasoning and decision-making (note that the title of the book indicates a rejection of the rationalist tradition of Descartes):

> I began writing this book to propose that reason may not be as pure as most of us think it is or wish it were, that emotions and feelings may not be intruders in the bastion of reason at all: they may be enmeshed in its networks, for worse *and* for better …. I suggest only that certain aspects of the process of emotion and feeling are indispensable for rationality. At their best, feelings point us in the proper direction, take us to the appropriate place in a decision-making space, where we may put the instruments of logic to good use. We are faced by uncertainty when we have to make a moral judgment, decide on the course of a personal relationship, choose some means to prevent our being penniless in old age, or plan for the life that lies ahead. Emotion and feeling, along with the

covert physiological machinery underlying them, assist us with the daunting task of predicting an uncertain future and planning our actions accordingly (Damasio 1994, pp. xii–xiii).

Damasio illustrates the role of emotion in decision-making with a patient he met:

> His practical reason was so impaired that it produced, in the wanderings of his daily life, a succession of mistakes, a perpetual violation of what would be considered socially appropriate and personally advantageous. He had had an entirely healthy mind until a neurological disease ravaged a specific sector of his brain and, from one day to the next, caused this profound defect in decision making. The instruments usually considered necessary and sufficient for rational behavior were intact in him... There was only one significant accompaniment to his decision-making failure: a marked alternation of the ability to experience feelings (Damasio, 1994, pp. xi–xii, 34–51).

A sophisticated philosophical defense of the role of emotions in ethics is found in Martha Nussbaum's book *Upheavals of Thought: The Intelligence of Emotions* (Nussbaum 2003). This is how Nussbaum summarizes her view of the significance of emotions in ethics:

> If emotions are suffused with intelligence and discernment, and if they contain in themselves an awareness of value or importance, they cannot, for example, easily be sidelined in accounts of ethical judgment, as so often they have been in the history of philosophy. Instead of viewing morality as a system of principles to be grasped by the detached intellect, and emotions as motivations that either support or subvert our choice to act according to principle, we will have to consider emotions as part and parcel of the system of ethical reasoning. We cannot plausibly omit them, once we acknowledge that emotions include in their content judgments that can be true or false, and good or bad guides to ethical choice .... To say that emotions should form a prominent part of the subject matter of moral philosophy is not to say that moral philosophy should give emotions a priviledged place of trust, or regard them as immune from rational criticism: for they may be no more reliable than any other set of entrenched beliefs .... It does mean, however, that we cannot ignore them, as so often moral philosophy has done (Nussbaum, 2003, pp. 1–2).

Of particular interest is Nussbaum's cognitivist account of emotions as including judgments that can be true or false, and her view that emotions can function as good or bad guides in ethical deliberation. It appears that on this cognitivist account emotions are quite similar to intuitions.

Let me also point out that Midgley's view of the complementary relation of reason and feeling fits very well the idea of moral imagination that I will present below. In his empirical investigations, the cognitive linguist Mark Johnson has shown that moral imagination—including empathy—rather than formal calculation characterizes moral reasoning (Johnson, 1993). These empirical findings should be taken seriously. I have tried to do so in my own ethical approach "imaginative casuistry" (Nordgren, 1998; Nordgren, 2001, pp. 15–49).

A major problem in accepting feelings or emotions as part of ethical deliberation is the fact that there may be many conflicting feelings or emotions within a person. This is precisely why ethics—according to the dominating view—should be a purely rational enterprise. One possible way of handling this within the alternative view that reason and feelings are supplementary could be to make a distinction between occasional feelings and more permanent deeper feelings. But even within the category of permanent deep feelings there could be different subcategories. We have seen in our analysis that Midgley makes a distinction between evolutionarily caused feelings and socially caused feelings. Examples of the former are speciesist and familyist feelings. Examples of the latter are racist and sexist feelings (Midgley, 1983, p. 96–97).

But why would evolutionarily caused feelings be more reliable ethically? Even evolutionarily caused feelings point in different directions. We cannot simply by means of our rational ability "read them off" and ascribe ethical weight to them. We need to use our rationality to balance the feelings from case to case.

A similar problem arises with regard to common feelings or intuitions regarding ethical issues in society. Within a particular society there may be many different and conflicting feelings or intuitions regarding, for example, the use of animals in research. This is why we cannot simply "read off" public opinion and use it as an ethical guide. As we saw in Chapter One, polls and surveys may indicate some common feelings but also significant differences. However, to the extent that common feelings exist, it might be justifiable to take these as a starting point in ethical deliberation. They must then be interpreted, balanced against each other, and related to already accepted ethical concerns. In the end, they might even be renounced as a result of critical evaluation. The situation is well described by Richard Hare, who distinguishes an intuitive and a critical level in moral thinking (Hare, 1981).

The conflicting feelings or intuitions within a single person and within a particular society make it necessary to use reason in an act of balancing. In the following two sections, I will discuss in more detail how to balance feelings of partiality and feelings of impartiality, and how to reason or argue from "is" to "ought" regarding facts about human nature including feelings.

6. Impartiality *and* Special Obligations

Behind the issue of antispeciesism versus speciesism, approached in different ways by the different prototypes, lies the more basic issue of impartiality versus special obligations. The latter issue is a classical problem in ethics (*cf.* LaFollette, 1993). Kant, representing the dominant line of thought, stresses impartiality. Ethics is an impartial enterprise. According to this established view, ethics is by definition impartial. To take on the ethical perspective is precisely to be impartial.

Hume, on the other hand, representing a minority strand of thought, stresses the importance of special obligations.

> A man naturally loves his children better than his nephews, his nephews better than his cousins, his cousins better than strangers, where everything else is equal. Hence arise our common measures of duty, in preferring one to the other. Our sense of duty always follows the common and natural course of our passions (Hume, 1978, pp. 483–484).

In stressing special obligations, Hume comes close to Aristotle. Writing on friendship, Aristotle argues that

> the duties of parents to children and those of brothers to each other are not the same, nor those of comrades and those of fellow citizens, and so too with other kinds of friendship. There is a difference, therefore, also between the acts that are unjust towards each of these classes of associates, and the injustice increases by being exhibited towards those who are friends in a fuller sense; e.g. it is a more terrible thing to defraud a comrade than a fellow citizen, more terrible not to help a brother than a stranger, and more terrible to wound a father than anyone else (*Nicomachean Ethics* viii.9, 1160a1–6).

Many attempts have been made to combine the idea of impartiality with the idea of special obligations (LaFollette, 1993). Hume suggests such a combination with regard to human beings (Hume, 1978).

It appears that Singer, Regan, Carruthers, and Cohen follow Kant (in this respect), while Midgley follows Hume. Singer and Regan attempt to be impartial in animal ethics by acknowledging animal interests and animal rights, respectively. Carruthers and Cohen can be interpreted as arguing that being impartial would lead us to accept that the idea of rights is possible only within a human community.

Midgley, on the other hand, stresses that we have special obligations due to social bonding. We have special obligations to our children compared to strangers, and we have special obligations to human beings compared to nonhuman animals. Taking our strong moral feelings of special obligations seriously is not prejudice, but a central aspect of being moral. From an evolutionary point of view, all species act for the benefit of their own kind over that of any other species and protect their own kin against unrelated members of the species. In the case of human beings, this should not be condemned as "speciesist" or "familyist" (*cf.* Hursthouse, 2000, pp. 127–132). However, we should always watch out for unjustly favoring our species or kin. Reason has the delicate task of discriminating between proper and improper feelings. A tension exists between protecting special claims and securing justice for outsiders. We must handle this tension by working out particular priorities in particular contexts (Midgley, 1983, p. 103).

This means that Midgley—like Hume—tries to combine impartiality and special obligations. But she does so not in the sense that impartiality would have us taking on special obligations like, for example, LaFollette (1993, p. 332) would have it, but in the sense that impartiality sometimes outweighs special obligations. Family and species preferences should be balanced against interhuman and interspecies justice.

It is interesting to note that Hume's description of natural preferences has received support from present-day empirical studies. Lewis Petrinovich has shown that human beings do have "intuitions" of stronger moral obligations to their own children than to strangers and to human beings compared to non-human animals. The research subjects were presented with hypothetical choice situations. These were of two types: "trolley problems" and "lifeboat problems." In the trolley problems, a decision is to be made whether or not to "throw the switch" that would determine which individual-or-group X or individual-or-group Y is killed. In the lifeboat problems, a decision is to be made to determine which among six members of a lifeboat survive. By focusing on extreme situations like these, Petrinovich believes that our deepest moral intuitions can be discovered (Petrinovich, 1998, pp. 151–176). He argues that these biases can be explained by the theory of evolution (Petrinovich, 1998, pp. 143–146). He also maintains that these empirical and theoretical findings are normatively relevant (Petrinovich, 1998, pp. 174, 206, 238; Petrinovich, 1999, pp. 3–4; see also Nordgren, 2002).

I agree with Petrinovich and I support Midgley's idea that we have special obligations due to social bonding. In discussing these issues, I find a couple of distinctions put forward by Brody very useful. He urges us to distinguish the question "Why should the interests of my children count more than do those of others?" from the question "Why should the interests of my children count more *for me* (my italics) than do those of others?" While the former has no answer, the latter has. Moreover, we should distinguish the question "Why should the interests of humans count more than do those of animals?" from the question "Why should the interests of humans count more *for human beings* (my italics) than do those of animals?" Also in this case the former question has no answer, while the latter has. The answer to the two answerable questions is that we have special obligations (Brody, 2001, pp. 143–144). However, it might be possible to accept that we have special obligations without accepting Midgley's particular argument for it, namely social bonding.

According to Singer, we may to some extent treat human beings differently from animals. Equal consideration does not necessarily imply equal treatment. Human beings may suffer more than animals in some respects and this is morally relevant. What Midgley affirms, but he denies, is that even when no quantitative difference in the amount of suffering exists, the human suffering counts more morally (*cf.* Brody, 2001, p. 142).

On at least one point it is not quite clear what Midgley's position is, namely whether a difference exists between our obligation to help individuals and our obligation to refrain from harming them. A reasonable interpretation would be that we commonly have a greater obligation to help the members of our family when they need it than to help a stranger, but when it comes to avoiding directly harming people we commonly do not have a greater obligation to avoid harming the members of our family than to avoid harming a stranger. However, with regard to animal experimentation there may be some exceptions. Our obligation to help human beings by developing new drugs may require that we harm laboratory animals in a way that we would not harm human beings (see below).

Let me quote one more participant in the animal ethics debate who argues in line with Midgley, namely Jerrold Tannenbaum.

> We do these things for our pets because we *care* about them .... It is not irrational or ethically indefensible to care about these animals while accepting the use of others, even members of the same species, in research. We are generally justified in heeding the needs and desires of members of our families more closely than we do those of strangers, and we have ethical duties to family members that sometimes require ignoring or even slighting others. Likewise, it is both sensible and sometimes ethically *obligatory* for us to care about and seek the health, welfare, and happiness of pets (Tannenbaum, 2001, p. 122).

The dominance of the principle of impartiality has recently been criticized also in ethical fields other than that of animal ethics. Brody states:

> I see no reasonable alternative for the adherent of the discounting position except to challenge the whole idea that we are, in general, morally committed to an equal consideration of interests. This is a plausible move, since equal consideration of interests has come under much challenge in contemporary moral philosophy, totally independently of the debate over the moral significance of the interests of animals (Brody, 2001, p. 142).

Recent examples of attempts to combine impartial and partial principles in ethics can be found in a special issue of the journal *Ethical Theory and Moral Practice*. In the introductory section "The Debate on Impartiality: An Introduction," the editor Albert W. Musschenga summarizes the contributions as follows:

> When is it appropriate to act on partial principles, and when on impartial ones? Most authors endorse the view that neither of the two types of principles always has priority over the other. Impartial principles are limited and corrected by partial principles, and vice versa. Which principle should get priority when a partial principle conflicts with an impartial principle depends on the nature of the situation and the stringency of the principles at stake (Musschenga, 2005, p. 8)

With this in mind, a practical key issue regarding animals becomes: when do special obligations to human beings outweigh interspecies justice and when does interspecies justice outweigh special obligations to human beings? As Midgley points out, we must learn to distinguish legitimate and illegitimate discounting of animal interests. The problem is how far social bonding takes us in discounting animal interests. Obviously not all the way to "strong human priority"! We need to balance obligations arising out of social bonding and those arising out of interspecies justice.

A related issue is whether it is possible to breed animals for scientific experimentation, farming, or companionship without "exploiting" them. It is one thing to protect and help animals that are in our care, quite another to breed them for "exploitation." For the sake of clarity we must distinguish two different meanings of "exploitation." In one sense the word means using animals as a means only. In another sense it means using animals as a means without caring for their welfare. I would argue that exploitation is ethically unacceptable in both senses. Using animals as a means for vital human welfare interests is acceptable if and only if we care for their welfare. Breeding them for such uses would not necessarily be exploitation in either sense. As Midgley points out, we should view it as something good and special for the human species to include other species in our community. We should support the destruction of human slavery, although we should not view breeding animals with care as animal slavery. We should view it as a creation of a mixed community of human beings and animals, ideally characterized by reciprocal benefit. However, there may arise situations of serious conflict. In these situations our obligations to human beings—due to social bonding—commonly outweigh our obligations to animals.

Finally, we have the problem of how far we should go in including animals in our community. One option is that we use pets as "prototypes" for how to include animals in our community and care for them. This means that we should try to find ways of, for example, farming that are in line with our care for pets. However, as pointed out by Tannenbaum, this approach creates problems with regard to animal experimentation:

> Ultimately, the most important effect of the emerging approach will be that many—perhaps most—research animals will be viewed in much the same way as we view pets. We will come to care about them and their lives so much that experimenting on them will be unthinkable (Tannenbaum, 2001, p. 121).

Tannenbaum is critical of this development. He argues: "This approach is dangerous precisely because its endorsement by people who are committed to using animals obscures the fact that it threatens animal research" (Tannenbaum, 2001, p. 93). He stresses that animal research is extremely important:

> The contributions of animal research to the health, safety, and well-being of both humans and animals have been enormous. Without animal research, very few of the medical advances we expect today for ourselves and our loved ones would be possible (Tannenbaum, 2001, p. 123)

I tend to agree with Tannenbaum. Animal experiments constitute difficult non-prototypical cases, deviating from the pet prototype. We should definitely work for their replacement, reduction, and refinement (Russell and Burch, 1992). However, for the foreseeable future it is difficult to see how they can be generally renounced.

7. From "Is" to "Ought"

With regard to all three issues above—intrinsic versus relational properties, reason versus feeling, and impartiality versus special obligations—I have argued that empirical studies provide important input to the discussion. The key question remains whether empirical studies are truely relevant for normative ethics.

It is quite common in ethics to start from natural properties. We see this in all the prototypes discussed above. Carruthers (1992) focuses on rationality and consciousness, Singer (1993a; 1995) on the ability to feel pleasure and pain (sentience) and on being a "person" with self-consciousness and rationality, Regan (1983) on the properties of a subject-of-a-life, and Cohen (1994) on moral ability. These are all intrinsic properties. Midgley (1983) on the other hand, focuses not only on intrinsic properties—mainly sentience—but also on relational properties due to social bonding.

The question is how to go from fact to value, from natural properties to moral properties, or from a description of "is" to a normative "ought." It is a common assumption in philosophy that an unbridgeable gap exists between "is" and "ought." The attempt to move from "is" to "ought" is sometimes called "the naturalistic fallacy" or "Hume's law." Neither designation is quite adequate, however. The term "naturalistic fallacy" is often reserved for the fallacy of defining values in terms of facts. The designation "Hume's law" is used to indicate that it was Hume who first discovered it (Hume, 1978, pp. 469–470). The fallacy is thought to be committed when we derive an "ought" from an "is" ("ought" is here a normative ought, not a descriptive ought). However, Larry Arnhart and others have shown that "Hume's law" is wrongly attributed to Hume (Arnhart, 1998, pp. 69–70; Buckle, 1991, pp. 282–284; Capaldi, 1989). In fact, Hume himself does derive an "ought" from an "is." He considers morality as rooted in the natural inclinations of human beings. A study of the context—the context within his book and the historical context—makes it clear that what Hume criticizes is the view that moral distinctions can be derived from abstract reasoning about structures in the universe that are completely independent of human nature, not the view that they can be grounded in human nature. Actually, the dichotomy of "is" and "ought" was first articulated by Kant who used it in a critical argument against Hume. Hume's thesis is that moral distinctions are derived not from pure reason alone but from a natural moral sense. Kant, on the other hand, treats morality as an autonomous realm. This realm is governed by its own internal logic with no reference to anything in human nature such as natural inclinations (Arnhart, 1998, pp. 69–83).

Let me discuss in more detail Midgley's suggestion that the feelings of social bonding—giving rise to special relations—are the result of evolution and something to be taken seriously in normative ethics.

The evolutionary part of Midgley's argument is somewhat underdeveloped in her book *Animals and Why They Matter*. She is more elaborate in *Beast and Man* written a few years earlier. She clarifies:

> The facts of evolution cannot guide us directly. They matter only insofar as they can help us to understand our nature, our emotional and rational constitution. Yet our understanding of that *does* give us practical guidance. Facts about it are directly relevant to values. Values register needs. It is a mistake to suppose that there is some logical barrier, convicting such thinking of a "naturalistic fallacy" .... We are not, and do not need to be, disembodied intellects. We are creatures of a definite species on this planet, and this shapes our values (Midgley, 1979, p. xxii).

Midgley concludes:

> If we say something is good or bad for human beings, we must take our species's actual needs and wants as facts, as something given. And the same would be true if we were speaking of any other species.... Moreover, our basic repertoire of wants is given .... We are not free to create or annihilate wants, either by private invention or by culture. Inventions and cultures group, reflect, guide, channel, and develop wants; they do not actually produce them. Thus if twentieth-century people want supersonic planes, they do so because of wants that they have in common with Eskimos and Bushmen ... We are innately "programmed" to want

and like such things ... The question is never which wants to have. It is always what to do about conflicts between existing ones (1979, pp. 182–183).

Midgley is right in stating that feelings, wants, and needs should be taken seriously, and that when they conflict, we have to prioritize between them. Midgley's idea that we are innately "programmed" needs clarification. If she defends a kind of genetic determinism, she is wrong, but it is possible and more reasonable to interpret her in a way that is more in line with modern behavioral genetics. All complex behavior is the result of interaction between genes and environment (*cf.* Nordgren, 2003; Parens *et al.*, 2006). Take, for example, the parental propensity of caring for biological offspring. This propensity is of key importance for reproductive success and likely a result of evolution. From the perspective of evolution, we therefore have reason to expect that the subject of parental feelings of care is primarily biological offspring. But parental feelings can be extended toward other subjects. Even non-human species may to a varying extent adopt a young animal of another species, although this is rare. This means that human beings may have parental feelings both to adopted human beings and to adopted animals, that is, pets.

I share Midgley's view on the is/ought issue and on the relevance of evolutionary theory for normative ethics. "Is" may be relevant for "ought." It is part of every moral problem situation to take a stand on what the relevant facts are. These facts may include characteristics of the specific situation and the historical and cultural context. But they may also include the biological context. Part of this context is the evolved human nature (*cf.* Nordgren, 2002). My arguments are as follows.

First, facts about our evolved human nature give us reason why we should be moral. Evolution has made us creatures that cannot flourish and lead a complete life if we do not act morally. We are social and moral animals. This is not just a matter of explanation but also of justification, although only a weak one. We have deep feelings that we should be moral. We desire justice and reciprocity. These feelings justify why we should be moral. Ultimately, we cannot justify rationally that we should be moral. We only feel that way. Psychopaths lack such moral feelings. Therefore we judge them abnormal. No more ultimate reasons can be given (except perhaps religious reasons). It is not possible to convince psychopaths with mere rational arguments why they should be moral. If we lack moral feelings, no further justification can be given.

Facts about our evolved human nature may also provide some intersubjectivity in ethics (Arnhart, 1998, pp. 1–13, 29–49). Nearly all human beings share some moral feelings and desires. This does not mean that they may not construct moral rules that may differ. On the contrary, the ways of balancing the moral feelings and desires may differ from one cultural context to another. No universal moral rules can be found, but (nearly) universal moral impulses appear to exist (Wilson, 1993, p. 18). These moral impulses make intersubjectivity possible to some extent.

Moreover, facts about our evolved human nature may provide content to our fundamental moral values. They inform our fundamental values by clarifying basic needs and desires. However—as indicated above—we cannot simply "read off" fundamental values from facts about human nature. As Janet Radcliffe Richards points out, "reading off" presumes a harmonious view of nature, which is inadequate after Darwin (Radcliffe Richards, 2000, pp. 246–247). No human "essence" exists, only more or less frequent characteristics. As stressed by Midgley, we have a bundle of innate tendencies and desires, and they may be in conflict with each other (Midgley, 1979, p. 183). This means that we cannot "derive" moral conclusions from statements about desires and inclinations in the sense of logical entailment. We have to justify the bridging of the is/ought gap by means of a value judgment based on practical wisdom. The evolved human nature should be affirmed, but not completely, neither should it be curbed completely. We must come to a wise balancing in each particular situation.

Furthermore, facts about our evolved human nature inform us when we are to implement our fundamental values by highlighting ethically unacceptable tendencies that we should counteract and difficult situations in which these tendencies might overwhelm us and which we should avoid (*cf.* Alcock, 2001, pp. 189–215).

Facts about our evolved human nature may also inform us when we are to implement our fundamental values by helping us develop ethical proposals that are practically feasible (Petrinovich, 1998, p. 38). "Ought" implies "can." If we do not take human nature into consideration, we run the risk that our proposals will not have any impact on the majority of people. A few individuals may accept our views, but they would appear too extreme or irrelevant to most people. Even those individuals that accept these views may find it difficult to keep them over an extended period of time. Our ethical proposals must be psychologically realistic. This is especially important in social ethics.

Finally, what does a reference to evolution add to the argument from social bonding? Midgley is not quite clear on this point. However, she makes a distinction between properties that are the result of cultural influences and those that are the result of evolution. The preference for our species is not "like race-prejudice, a product of culture. It is found in all human cultures, and in cases of real competition it tends to operate very strongly" (Midgley, 1983, p. 104). In other words, our speciesist propensities are very difficult—if not practically impossible—to get rid of, since they are the result of evolution rather than cultural influences. They are part of who we are. We are not abstract intellects but creatures of flesh and blood that are the results of millions of years of evolution. We have to accept speciesist propensities as a fact of human nature precisely because of their evolutionary background. But they are not the only propensities that are the result of evolution. Also other propensities exist, tending in another direction, namely propensities for sympathy for animals and justice for animals. As already mentioned, these evolutionary tendencies may be in conflict, and it requires wisdom to balance them in concrete cases. Thus, a reference to evolution can add only weak support to the argument from social bonding.

Let me also briefly comment on Midgley's statement that race-prejudice is a product of culture. We do not know this for sure. It might be possible that such prejudice, one way or another, has evolutionary roots. But even if this were the case, that would not mean that it should be accepted. We have also an inborn sense of justice, and this sense would have us reject this prejudice. As already pointed out, we have a bundle of innate tendencies, and these may be in conflict with each other. And it is possible to give priority to our children over strangers—in most cases of conflict—without accepting race-prejudice.

With these considerations in mind, we can draw the conclusion that it is possible to bridge the gap between "is" and "ought," although it requires wisdom to do so. The bridging of the is/ought gap is not primarily a matter of formal validity but of substantial soundness (*cf.* Toulmin, 1958, pp. 94–145). The real problem is not whether it is logically possible to derive an "ought"-statement from an "is"-statement. This would formally require another statement—for example, "If 'is', so 'ought'"—functioning as a bridge between these two statements. The real problem concerns which "is"-statements are relevant for which "ought"-statements in which contexts, and this is a matter of judgment rather than formal logic. As I interpret Midgley, it is not always justified to build a bridge between the statement "It is part of our evolved human nature to feel stronger moral obligations to human beings than to animals" and the conclusion "We should accept that we have feelings of stronger moral obligations to human beings than to animals and we should act accordingly." We can draw that conclusion in many cases, but not always. It may sometimes be reasonable to let our feelings of sympathy with animals and of interspecies justice have more weight. Conflicts between innate tendencies often occur. Not everything in human nature is always acceptable. This is why wisdom is necessary to achieve a reasonable balance.

How is all this relevant to animal experimentation? Facts about our evolved human nature may inform our ethical views on animal experimentation. Feelings of social bonding with human beings may be referred to in order to support the ethical view that we have stronger moral obligations to human beings than to animals, leading to a weak normative speciesism. Feelings of sympathy and social bonding with animals may be referred to in order to support the ethical view that we should care for animals, leading to an obligation of interspecies justice. However, weak normative speciesism and interspecies justice may sometimes be in conflict. Animal experimentation involves such a conflict. Practical wisdom is required to balance it. The outcome of such balancing will probably be that vital human research interests outweigh animal harm. Sometimes, however, animal harm may outweigh human research interests. In this way, we may develop a psychologically realistic social ethic of animal experimentation on the basis of facts about our evolved human nature.

8. Strengths and Weaknesses of the Five Prototypes

As is obvious from this discussion, I agree with Midgley on four basic presumptions.

(1) Not only intrinsic properties but also relational ones may be ethically relevant.
(2) Feelings are important alongside reason in ethical deliberation.
(3) Impartiality and special obligations are to be balanced.
(4) It may be acceptable to argue from "is" to "ought."

Given these presumptions, let me briefly assess the strengths and weaknesses of the five prototypes.

A strong aspect of Carruthers's (1992) view is that it takes seriously the common moral intuition that we have stronger moral obligations to human beings than to animals. A weakness is that he is categorical on this and does not accept that we have direct duties to animals, and he is obviously wrong in his tentative denial of animal consciousness (see Chaper 5).

Cohen (1994)—together with Carruthers—is partly correct in criticizing Regan's view that animals have rights, but he does not acknowledge the possibility of distinguishing rights in a strong sense from rights in a weak sense. It is probably not possible to attribute rights in the strong sense—as correlates to having duties—to animals, although they may be attributed rights in the weak sense—without being correlates to having duties. A strength compared to Carruthers is that Cohen maintains that we have direct duties to sentient animals and that we should always try to minimize their suffering. A weakness with Cohen is that he does not acknowledge that some animal experiments should not be carried out at all because the cost for the animals is too high.

A strength of Regan's (1983) view is that he acknowledges the inherent value of animals and attempts to counteract injustice to animals by stressing animal rights. This means that the burden of proof lies on those who want to carry out animal experimentation. A weakness is that he is not prepared to balance animal rights and expected human benefit in such experimentation. He is also too pessimistic about the benefits of animal experimentation.

Singer (1993a; 1995) is correct in taking animal suffering very seriously, but he is wrong in his categorical criticism of speciesism. A weak normative speciesism might be ethically acceptable. He is also wrong in his very negative view of most animal experimentation.

Midgley's (1983) view encompasses all the strong aspects of the other views, while avoiding the weaknesses. She takes the moral intuition seriously that we have strong obligations to animals—as Singer and Regan do—but she does so less radically. She accepts the moral intuition—stressed by Carruthers and Cohen—that we have stronger moral obligations to human beings than to animals, but does so less categorically. We commonly—but not always—have stronger obligations to human beings than to animals. Not all animal experiments are acceptable. We have to balance expected human benefit and animal suffering on a case-by-case basis.

9. Proposal: Weak Human Priority

Given the discussion of basic presumptions and the above assessment of strengths and weaknesses of the prototypes, it should have become quite obvious that I come close to Midgley's weak human priority prototype. However, important differences exist. Below I will give a more developed argument for the weak human priority view as a general position. I will also clarify in which respects my version of this view differs from Midgley's and also give reasons for why my version is preferable.

A. Animal Experiments Are Wrong, Unless ...

The starting point of my version of the weak human priority position is that experimentation that inflicts harm on animals is *prima facie* wrong. This is not explicitly maintained by Midgley, although it is compatible with her view. Animal experiments are ethically unacceptable, *unless* certain special considerations suggest the opposite.

Three different arguments suggest why animal experimentation is *prima facie* wrong. They are all non-relational and take some aspects of Singer's and Regan's main points seriously.

The first argument is the argument from infliction of animal pain. This argument takes animal sentience seriously. Singer is correct in stressing this aspect. However, in my version the argument does not imply that all or almost all animal experiments are wrong. Instead it suggests that if we do not have strong special reasons, we should not carry out experiments that inflict pain on animals. In addition, the argument suggests that an experiment that involves more animal pain is worse than one that involves less.

A second argument is the argument from violation of animal integrity. This argument is closely related to—but not identical with—Regan's argument from the inherent value of animals. The word "integrity" derives from the Latin word *integritas*, which means "untouchedness, wholeness." Violation of integrity is a metaphor that illustrates how something that is an untouched whole is touched and thereby broken into pieces. A clarifying analysis of the concept of animal integrity has been made by Bart Rutgers and Robert Heeger. They argue that in a state of integrity the following three elements must be present:

(1) the wholeness and completeness of the individual animal,
(2) the species-specific balance of the creature,
(3) the animal's capacity to maintain itself independently in an environment suitable for the species (Rutgers and Heeger, 1999; Heeger 1997).

Animal experimentation may imply a violation of animal integrity in all these respects. This is especially clear when animals are used as disease models or when toxic substances are tested on them. In both cases, the balance of the animals is disturbed and their ability to maintain themselves is reduced. Moreover, killing—at the end of the experiment—is a violation of animal integrity, since the whole is completely broken into pieces and ceases to exist.

My version of this argument is non-categorical. Animal experiments are ethically unacceptable, unless special considerations carry more weight. Animal integrity suggests that the burden of proof lies with those who want to carry out animal experimentation. The violation of integrity represents a cost to be taken into account in any ethical balancing regarding animal experimentation in addition to the pain that is inflicted. Not only animals that are subjects-of-a-life have integrity in this sense, but also less developed animals. According to my version of the argument, the cost is higher when a more developed animal's integrity is violated. Its organismic integration is more complex.

A third argument is the argument from interspecies justice, and it is based on the first two arguments. It stresses that it is unjust to harm animals in order to do good to human beings. The harm can be conceptualized in terms of pain (in line with the first argument) or in terms of violation of integrity (in line with the second argument). Justice appeals to impartiality and speaks against favoring our species at the expense of another. In my version, the argument is non-categorical. It sets some limits for the human use of animals in experiments. Animal experiments should not be carried out, unless we have special reasons to do so.

B. The Argument from Species Care

Thus, we have three strong reasons to consider animal experimentation *prima facie* wrong. Animal experimentation is wrong *unless* certain conditions are satisfied. So, under what conditions is animal experimentation acceptable? We have seen that Midgley's argument for weak human priority is relational and that it stresses the ethical relevance of the social bonding between human beings. It is still not quite clear in what way and to what extent social bonding justifies animal experimentation. In order to clarify this, I would like to refer to an article by Jennifer Welchman. She discusses ethical aspects of xenotransplantation, but I will explore the implications of this type of argument for animal experimentation in general (Welchman, 2003).

Welchman distinguishes two different relational arguments for prioritizing human beings over animals, the argument from species loyalty and the argument from species solidarity.

As an example of an argument from species loyalty, Welchman quotes Stephen Post who maintains that

> our species kinship is a family-like phenomenon creating special obligations of beneficence that take precedence over our obligation to members of other species (Post 1993, p. 295).

He also points out:

> While justice requires stricter impartiality, obligations of beneficence allow for considerable partiality. One can accept the idea that we should extend our beneficence to humans before nonhumans because of our species loyalty, but deny that this ... allows us the liberty to inflict cruelty on animals (Post, 1993, p. 295).

Welchman criticizes Post's notion of species loyalty. Loyalty is based on common social activities and we do not stand in such a close relation to our species as a whole (Welchman, 2003).

As an alternative, Welchman suggests an argument from species solidarity. Solidarity does not imply that we know all those that we feel solidarity with. Therefore, a reference to solidarity is preferable to a reference to loyalty when applied to the human species (Welchman, 2003). She states:

> Thus an animal's innocence of any direct threat to human life does not entail that a human must treat its interests impartially with the interests of those with whom she is in solidarity (i.e., human beings), if those interests conflict (Welchman, 2003, p. 251).

She concludes:

> Human solidarity can, *in certain circumstances*, morally justify decisions to use animals rather than fetal or mentally disabled humans as sources of organs for other humans. However, the argument I have considered does *not* support the further claim that people in general or biomedical researchers in particular have an obligation to harm animals in order to assist persons. Appeals to human solidarity are, after all, only a special case of appeals to the right of self-defense (Welchman, 2003, p. 253).

Several things should be noted here. First, solidarity suggests a common enemy, even if we do not know everybody that is threatened by this enemy. However, our enemies are not the animals but different life-threatening diseases. Second, Welchman is defending the use of animal organs for xenotransplantation, not animal experimentation in general. Third, she believes that using animal organs for xenotransplantation is acceptable only for saving lives, not for treating conditions that are not life-threatening.

I agree with Welchman that the argument from species solidarity is preferable to the argument from species loyalty, but I think that an even better metaphorical framing exists, namely species care. The notion of care has a more open scope than loyalty, since it does not require that you know all those that you care for. But it is still more family-like than solidarity, since the prototype of care is the care for our children. More than others Hans Jonas has stressed the parent-child relation as the "archetype of responsibility" (Jonas, 1984, pp. 130–135). The clearest example of responsibility concerns

> the newborn, whose mere breathing uncontradictably addresses an ought to the world around, namely, to take care of him (Jonas, 1984, p. 131).

The force of parents' commitment to their children's health and welfare is typically extremely strong. Most parents would be prepared to do almost anything to help their children if they get a disease, in particular if it is life-threatening. In the argument from species care, this care for our children is extended to the whole human species, from our children to all children, and from our family to "the human family." The reference to "our children" (in the literal sense) in this argument has an obvious limitation. If our children have already got a disease or will get a disease quite soon, a new cure or treatment may not be available in time to help them, despite intensive research on animals and human beings. It commonly takes 10 to 15 years to develop a pharmaceutical drug, and then the time for basic research is not included. This means that the argument as far as it refers to "our children" primarily points out a prototype of care that should be extended to the whole patient group, that is, to all children and/or adults sharing this particular disease or condition.

Care for our children, whether biological or adopted, for all children, and for the whole human species justifies that we, on certain conditions, carry out experiments on animals as an important step in the development of medical treatments for life-threatening diseases. But I also argue that species care justifies animal experiments aiming at finding treatments for conditions that are not life-threatening but involve severe or moderate pain or suffering, including experiments within basic research providing the basis for such development of medical treatments. At this point, I differ from Welchman.

For the sake of clarity, let me also stress again the distinctions by Brody mentioned earlier. He distinguishes the question "Why should the interests of my children count more than do those of others?" from the question "Why should the interests of my children count more *for me* (my italics) than do those of others?" While the former has no answer, the latter has. Moreover, he distinguishes the question "Why should the interests of humans count more than do those of animals?" from the question "Why should the interests of humans count more *for human beings* (my italics) than do those of animals?" Also in this case the former question has no answer, while the latter has (Brody, 2001, pp. 143-144). My answer to the two answerable questions is that we have special responsibilities due to special relations. The interests of our children should count more for us than those of others, because it is our parental moral responsibility to care more for them. And the interests of human beings should count more for human beings than those of animals, because it is our human moral responsibility to care more for them. Not caring more for our children and for our species would be irresponsible (for an analysis of the concept of responsibility, see Nordgren, 2001).

C. Objection and Response

However, the problem is how far this argument from species care takes us. A serious objection is that it is one thing to prioritize human beings over animals in doing good; it is quite another to harm animals in order to do good to human beings (Bernstein, 2004; Zamir, 2006; Zamir, 2007). Midgley and Brody do not explicitly recognize this distinction, although they think that it is ethically acceptable to harm animals in experiments in order to do good to human beings, at least to some extent. This is implied in their idea of partial discounting of animal interests. Tzachi Zamir, however, uses this distinction in an objection to this kind of an argument in favor of animal experimentation. He accepts the kind of speciesism that implies that we give human beings priority over animals in doing good, but objects to the kind of speciesism that involves harming animals in order to help human beings. As support for this, he refers to the analogy of helping citizens of our country before helping citizens of other countries. He finds this acceptable, but stresses that it is commonly agreed that helping citizens of our country by harming citizens of other countries is wrong (Zamir, 2006).

My response to this objection focuses on the prototype of care, namely the care for our children. If my child has a serious disease and I could possibly help her by inflicting minor or moderate pain on mice or rats—the prototypes of laboratory animals—this would be ethically acceptable. The moral responsibility of caring for my child outweighs the responsibility of not inflicting minor or moderate pain on mice or rats. This is not at all like harming citizens of other countries, as in Zamir's analogy. A fundamental disanalogy exists between harming mice and rats in experiments and harming citizens in other countries. The difference is that our relations to human beings differ from our relations to mice and rats. This relational difference confers different obligations to human beings compared to mice and rats. We should not harm human beings in other countries precisely because we stand in a human-to-human relation to them. We might, however, to some extent harm mice and rats because we do not stand in a human-to-human relation to them.

Zamir may be right that it is not possible to provide an analogy from the human context showing that it might be acceptable, in certain circumstances, to harm others in order to do good to those more closely related. However, I question the premise of this argument, namely that it is necessary to provide such an analogy in order to justify animal experimentation involving harm to animals. This might not be necessary. As the prototype of care—the care for our children—indicates, human-to-human relations might be special and impossible to provide any analogy for.

D. Balancing Species Care and Interspecies Justice

The argument from species care is not an argument for the ethical acceptability of all animal experiments (strong human priority), but only for some (weak human priority). The reason is that species care (a partial obligation) needs to be balanced against interspecies justice (an impartial obligation). This sets a limit for the use of animals in experimentation: an animal experiment is not ethically acceptable if the expected human benefit of the experiment is very low and the expected animal harm is severe. This implies that the long-term goal should be to stop carrying out animal experiments, and that we need to do much more do find non-animal alternatives. At present and for the foreseeable future, however, it would be ethically irresponsible to stop carrying out animal experiments. The care for our children and other human beings requires that we continue doing at least some animal experimentation.

E. Conditions

We have seen that experiments that inflict harm on animals are *prima facie* wrong. They should not be carried out, *unless* certain conditions are satisfied. Conversely, an animal experiment is ethically acceptable if these conditions are satisfied. The conditions are as follows.

(1) The purpose of the experiment is of vital human interest.
(2) The experiment is likely to be of human benefit.
(3) No non-animal alternatives are available in attaining the purpose of the experiment.
(4) The number of animals used is kept as low as possible given the purpose of the experiment.
(5) The experiment involves animals with as low a degree of sentience as possible given the purpose of the experiment.
(6) The experiment inflicts as little harm on the animals as possible given the purpose of the experiment.
(7) The harm that is inflicted on the animals is outweighed by the expected human benefit.

The first condition concerns the scientific purpose. The problem is which purposes are important enough to outweigh animal harm. The purpose of developing medical treatments for life-threatening diseases is a prototypical case, but also the purpose of finding treatments for conditions that are not life-threatening but involve severe or moderate pain or suffering. I would also include the purpose of providing a scientific basis for the development of medical treatments through basic research.

The second condition is that the human benefit that can be expected from carrying out the experiment has some likelihood. To make a judgment of the likelihood of benefit of a particular experiment is, however, quite difficult. In Chapter Five, I will clarify this problem and also provide a couple of suggestions on how to handle it.

Conditions (3)–(6) correspond to the 3Rs—replacement, reduction, and refinement—suggested by W. H. S. Russell and R. L. Burch (1992). The third condition corresponds to replacement, the fourth to reduction, and the fifth and sixth to refinement. In Chapter Four, I will discuss the 3Rs in more detail. I interpret the 3Rs, not only as goals to strive for, but as conditions that must be satisfied, and I view them as *ethical* conditions, although they are not presented as such in the book by Russell and Burch.

The third condition is that the animal experiment is necessary in the sense that no alternative non-animal methods exist for obtaining the scientific knowledge needed for developing the medical treatment in question. This means that the scientific value of animal experimentation is an extremely important issue. It becomes crucial to establish that it is possible to extrapolate results from animal experiments to human beings. Most people would probably agree in principle that to the extent animal experiments can be expected to contribute to the development of medical treatments these experiments are ethically acceptable, although some would argue that this is never the case or only very seldom (*cf.* Regan and Singer, respectively). For my argument it will therefore be crucial to show the scientific value of animal experimentation. I will take up this task in Chapter Four. Midgley is not clear on this point, and this issue is in need of further elaboration.

The fourth condition is also vital. Since we have the *prima facie* duty not to inflict pain or violate the integrity of animals, we should use as few animals as possible, given the scientific requirements.

It is especially important to note the fifth condition, which concerns animal species. The argument from species care does not justify that we carry out experiments on any species. We should experiment on animals with as low a level of sentience as possible given the scientific question we are trying to answer. This means that with regard to mammals primarily mice and rats are to be used, and never or almost never apes.

The sixth condition also requires an explanation. Certain limits exist regarding the degree of harm we can inflict on laboratory animals in order to help members of our human species. These limits are related to the balancing pointed out in the seventh condition. But what is harm? One aspect is pain and suffering. This is related to the concept of animal welfare. In animal experimentation, harm in this sense may range from minor to moderate and even severe. Another aspect of harm is violation of integrity. As mentioned above, being killed—even immediately and without pain—would be an example of harm in this sense.

The seventh condition concerns balancing of expected animal harm and expected human benefit. The expected human benefit must outweigh the expected animal harm. If the expected human benefit is very low and the expected animal harm is very severe, then the experiment should not be carried out. If, on the other hand, the expected human benefit is very high and the expected animal harm is minor, then the experiment is ethically acceptable. In between these two extremes, there may be difficult cases that require serious ethical deliberation. It is a very difficult problem exactly where to draw the line. What about experiments in which the expected human benefit is only moderate and the animal pain is moderate? In Chapter Five, I will discuss ethical balancing in more detail and also propose a simple matrix model for practical use.

F. Practical Implications

What would be the consequences in practice if this ethical view on animal experimentation—the weak human priority position—were to be implemented? The special obligations to our children and other human beings to find medical treatments can be expected to commonly—but not always—outweigh our obligations to animals. The likely result would be quite similar to the present-day assessments by animal welfare agencies and committees in some countries. The policy would probably be more restrictive than the policies of the United States and even the European Union, but more in line with, for example, the policy of my own country, Sweden (see the first section of this chapter). It would be more critical to animal experimentation than the pharmaceutical industry and many researchers but still more positive than many animal ethicists such as Regan, Singer, and Zamir. In addition, the weak human priority position would suggest that much more should be done to find non-animal alternatives than we see today.

10. Differences Compared to Midgley's Version

I will now summarize the main differences between my version of weak human priority and Midgley's.

A. Interspecies Justice and Species Care

As we have seen, I stress—in contrast to Midgley—that animal experimentation is *prima facie* wrong. It is not ethically justified, unless certain special considerations suggest the opposite. Moreover, my version of the social bonding argument is more developed than Midgley's and takes into account recent criticism.

B. Moral Imagination

Midgley emphasizes that reason and feeling are complementary in ethics. This is good, but one aspect is lacking, namely moral imagination. This holds true also of the academic debate on ethics in general, which has focused on logical reasoning rather than moral imagination. In my view, moral imagination is crucial in developing a well-considered normative ethics. It is also crucial for the ethics of animal experimentation. Below I will clarify what moral imagination is and what it implies.

C. Imaginative Casuistry

We have seen that Midgley shows a casuistic and pluralistic tendency, although she does not label her view as casuistic. She merely stresses that a balancing of a plurality of values should be made on a case-by-case basis. I have elsewhere developed and explicitly defended a casuistic view that I call "imaginative casuistry" (see Nordgren, 1998; Nordgren, 2001, pp. 15–49). The designation is due to the focus on moral imagination, mentioned above. Below I will present this view in more detail.

Let me clarify that a casuistic approach does not exclude ethical principles or rules. What is crucial is that these are ultimately not justified deductively from a more general ethical theory but inductively from paradigmatic or prototypical cases (see below).

Let me also point out that it is quite possible to combine a focus on public policy with a casuistic approach (*cf.* Nordgren, 2001, pp. 33–37). The plurality of values expressed in a particular public policy or legal regulation can be viewed as a starting-point, but they must always be balanced from case to case. This holds true in particular for the assessment of animal experiments.

D. The Scientific Value of Animal Experimentation

My argument in favor of a weak human priority position on animal experimentation makes the scientific value of animal experimentation an extremely important issue. It becomes crucial to establish that it is possible to extrapolate results from animal experiments to human beings. Midgley is not clear on this point. In the next chapter, I will discuss this issue fairly extensively.

E. A Comprehensive View of Animal Welfare

Midgley does not explicitly discuss the concept of animal welfare, but appears to understand it primarily in terms of feelings of pleasure (and poor welfare in terms of pain and suffering) or in terms of interests (Midgley, 1983, p. 96). Feeling well is an important aspect of animal welfare, but there might also be other aspects. An intensive discussion is going on in animal welfare science about the concept of animal welfare. In addition to feeling-based conceptions, there have also been suggested function-based conceptions and conceptions in terms of natural living. Some commentators suggest a combination of such conceptions. In Chapter Five, I will propose a comprehensive approach to animal welfare and discuss conceptual and ethical problems that this approach raises. I will also discuss animal welfare with a special focus on animal experimentation.

F. Ethical Balancing

Midgley stresses the aspect of balancing, but she does not clarify in more detail what it implies. Ethical balancing is central to the weak human priority prototype. In assessing an animal experiment, the expected human benefit and expected animal harm are to be balanced against each other. Balancing is also central to imaginative casuistry more generally. In Chapter Five, I will discuss different methods of balancing and also suggest a particular method to be used in the assessment of animal experiments in the public sphere.

11. Moral Imagination and Imaginative Casuistry

After this presentation of my version of the weak human priority position and my arguments, I will elaborate on some key aspects of this position in more detail.

Let me start by explicating my general ethical theory—imaginative casuistry—and in particular the role of moral imagination.

A. Empirical Findings in Cognitive Semantics

I have argued that empirical findings may be ethically relevant. One important finding made by the cognitive linguist Mark Johnson concerns the role of moral imagination (Johnson, 1993). Johnson found that in ethical deliberation moral imagination rather than formal calculation is used. The reason for this is that formal calculation requires literal and well-defined concepts, whereas moral concepts are in no way similar to this; they are metaphorical and exhibit a prototype structure (Johnson, 1993). Let me explain.

Above I mentioned the definition of "metaphor" as a concept from one domain of experience (the source domain) that is used to structure our understanding of another domain (the target domain) (Lakoff and Johnson, 1980; Lakoff and Johnson, 1999). Moral concepts such as "rights" and "obligations" are also metaphorical. Rights and obligations are basically financial metaphors of credit and debt, respectively (Johnson, 1993, pp. 35–50).

Moral concepts have prototype structure in the sense that they have prototypical instances at the center with non-prototypical instances radiating out at different distances. Prototypical cases are those that are clear and accepted by most people—in the sense that most people accept that they belong to this category—while non-prototypical cases are unclear and disputed—in the sense that some people do not accept that they belong to this category (Rosch and Lloyd, 1978; Lakoff and Johnson, 1999; Johnson, 1993, pp. 92, 189–192). Take, for example, the concept of coercion. A prototypical instance is the use of physical force to make a person do what you want. Some people may want to extend the concept also to the use of social pressure, economic pressure, or oral persuasion. Different views exist on just how far the concept of coercion should be extended.

Moral imagination can be defined as the ability to envisage alternative perspectives and arguments, to empathize with those affected by our actions, and to use moral metaphors and extend them to non-prototypical cases with discretion (*cf.* Johnson, 1993, pp. 198–203). Moral imagination prompts us to ask questions like: What metaphors should we use? How far should concepts be extended? With whom should we empathize? These questions can be answered only with the use of moral imagination; logical reasoning is not sufficient.

B. A Normative Approach Based on the Empirical Findings: Imaginative Casuistry

On the basis of Johnson's empirical findings, I have elsewhere developed an approach that I call "imaginative casuistry." Its key characteristics are moral imagination, a plurality of values and norms, and case-by-case balancing (Nordgren, 1998; Nordgren, 2001, pp. 15–49).

Let me explain in more detail what imaginative casuistry implies and what it does not imply. Imaginative casuistry provides a method of justification. Prototypical cases give content to ethical principles and provide their ultimate justification. The method is basically inductive (bottom-up). This does not mean that in everyday moral reasoning we always or even commonly go from particulars to general principles. It might very well be the opposite. Frequency and ultimate justification are two different things (*cf.* Winkler, 1993, pp. 361–362). We frequently use rules of thumb in practical decision-making, but ultimately these rules of thumb get their meaning and justification from prototypical cases. The prototypical cases provide the interpretation, scope of application, and relative weight of the rules of thumb (Nordgren, 2001, pp. 34–37). This means that the method of justification of imaginative casuistry differs from deductive approaches (top-down from theories or principles to particular cases) and coherentist approaches (for example, the reflective equilibrium defended by authors like Rawls (1971) and Daniels (1979); see also Beauchamp and Childress in their latest edition of *Principles of Biomedical Ethics* (2009, pp. 381–387).

Imaginative casuistry is not a view that provides clear-cut answers to difficult ethical questions. It is a view that takes seriously the complexity of ethical problems. In doing so it acknowledges a plurality of legitimate moral appeals, a plurality of values and norms (*cf.* Brody 1998; Strong 1997). This distinguishes it from approaches that are single-valued, that is, approaches that include only one single basic value or norm and derive secondary values or norms from this single value or norm.

The reason imaginative casuistry has this pluralistic character is the fact that it takes moral imagination as its methodological starting point. Using our moral imagination means envisioning several alternative ethical perspectives, and recognizing their valid points and their weaknesses with regard to their application to particular cases. We see something valid in many different ethical approaches such as utilitarianism, Kantianism, rights-based ethics, communitarian ethics, virtue ethics, and relational ethics. These approaches fail because they focus on only one of many legitimate moral appeals. Therefore, the legitimacy of each approach is limited. This explains why ethical problems have the character of balancing moral appeals, which in themselves are completely legitimate but may be in conflict with each other in particular cases. Many moral appeals are valid *prima facie*, but we have to determine in particular cases which one is actually valid.

Let me give a few examples of values and norms acknowledged *prima facie* by imaginative casuistry. By "value" I mean a good state of affairs. By "norm" I mean a principle or rule prescribing an action or type of action. Values and norms are often closely related in that a norm may prescribe that a value should be realized by an action or type of action. In human ethics—that is, ethics regarding how to act toward other human beings—important values/norms are respect for human dignity, respect for autonomy, individual rights, non-maleficence, beneficence, utility, justice, solidarity, precaution, virtues such as sympathy, and obligations based on special relations, for example, care for our children. The interpretation, scope of application, and relative weight of these values/norms may vary. Take the example of abortion. Differences in these respects with regard to the principle of respect for human dignity may lead to a variety of different views.

Conflicts between different values/norms are to be expected from the perspective of imaginative casuistry. These conflicts necessitate ethical balancing. In contrast to some pluralistic approaches, which imply a lexical balancing, that is, a hierarchical ranking always to be followed, imaginative casuistry implies a contextual balancing, that is, a balancing from case to case depending on the particularities of each case (*cf.* Strong, 1997).

C. Arguments, Objections, and Responses

Let me summarize the two main arguments in favor of imaginative casuistry. One argument is that imaginative casuistry takes the empirical findings of cognitive semantics seriously, which is not the case with the other approaches, at least not in their traditional versions. If metaphors are abundant in moral reasoning, if basic moral concepts have prototype structure, and if moral imagination rather than formal logic is central in moral deliberation, then these empirical findings should be acknowledged explicitly and incorporated also into a normative ethical framework (*cf.* Nordgren, 2001).

Another argument is that each of a plurality of moral appeals appears intuitively acceptable. Above I have given several examples of values and norms in human ethics and animal ethics that most people probably accept in one way or another. Imaginative casuistry takes the ethical experiences and insights of each approach seriously and attempts to include all these different values and norms; hence, its pluralistic character.

I will also briefly respond to some possible objections to imaginative casuistry. One objection is that it is theoretically unsatisfying. Values and norms have to be related in a coherent way, preferably by reducing the number of basic values and norms from which other values and norms may be derived and by avoiding or at least reducing conflicts between these basic values and norms. A response to this objection is that far from being theoretically unsatisfying, it is instead a theoretical strength of imaginative casuistry that it accommodates complexity in ethics by allowing a plurality of values and norms and by acknowledging conflicts between these values and norms in particular contexts. However, the objection is valid to some extent. If possible, a simpler theory is preferable to a more complex one, but we should not try to achieve simplicity by neglecting complexity. If it is not possible to reduce the number of basic values and norms, and if it is not possible to avoid or reduce conflicts between these values and norms, we should not try to do it. The result would be an over-simplified theory. The plurality should be accepted, and the conflicts should be resolved by contextual ordering, that is, on a case-by-case basis. However, it is sometimes desirable and possible to formulate simple coherent general policies covering several cases or types of cases. Examples are the Helsinki Declaration (World Medical Association, 1964, with later revisions) and UNESCOS's Universal Declaration on Bioethics and Human Rights (UNESCO, 2006). Imaginative casuistry does not exclude such guidelines, but it recognizes their limitations. They are often too abstract. In actual biomedical research, problems often arise in a gray zone not covered by the guidelines, requiring a balancing of conflicting values and norms.

Another objection is that imaginative casuistry runs the risk of being arbitrary. You can defend any position on any issue in this way. The response to this is that imaginative casuistry is context-sensitive instead of arbitrary. It is precisely this context-sensitivity that many of the traditional ethical approaches lack. However, the objection is valid to some extent. A risk of arbitrariness does exist, and we should always be aware of this risk and try to avoid it. Moral imagination offers a "transperspectivity" that helps us in this regard (*cf.* Johnson, 1993, pp. 240–243). Precisely by envisioning many alternative perspectives on an issue arbitrariness is avoided. This is not a "God's-eye-view" objectivity but a kind of "weak objectivity" that is possible for evolved human animals like us.

A related objection is that by accepting contextual factors imaginative casuistry might lead to complete normative relativism. We might end up with the conclusion that the deep cultural differences in the world today with regard to ethics should not be eliminated but accepted. Imaginative casuistry may help us also in this regard. The option of "transperspectivity" opened up by moral imagination makes it possible to understand other cultures, at least to some extent. It also makes it possible to find creative ways of establishing an "overlapping consensus" (Rawls, 1993) at the level of global policy, by justifying the same ethical principles or rules from different cultural perspectives.

My conclusion in this brief discussion is that imaginative casuistry is a tenable position (for a more extensive discussion, see Nordgren 1998 and 2001). Let us investigate its implications for the ethics of animal experimentation.

12. Moral Imagination in Animal Experimentation

In the ethics of animal experimentation, moral imagination suggests that we ask questions about metaphors such as: Is the metaphor of "rights" applicable to animals, that is, do animals have rights (Regan, 1983)? Is the metaphor of "obligations" applicable in relation to animals, that is, do we have obligations to animals? Actually, this aspect of moral imagination was the reason I analyzed key metaphors of the five ethical prototypes of animal experimentation in the beginning of this chapter.

Moral imagination also forces us to put animal experimentation into a broader framework. Which human interests—if any—are vital enough to outweigh animal suffering? Can animal interests sometimes outweigh the human interest in carrying out animal experiments? What are the benefits of particular experiments for future patients? To what extent is basic research, involving animals, ethically acceptable? What are the social consequences of undertaking versus not undertaking animal experimentation (*cf.* Nordgren, 2004)?

Moral imagination prompts us to ask what should be done if our children or grandchildren would suffer from this or that disease. This provides the proper starting point for ethical deliberation on animal experimentation in biomedicine. From this perspective, we can expand the circle and add the more general question: what should be done if other human beings suffer from this or that disease?

Moral imagination suggests that we ask how animals would experience animal experimentation, whether we should empathize with these animals, and what this would imply. An important issue in this context is whether it is possible to imagine what it is like to be a laboratory animal. Singer invites us to try to "imagine what it is like to be a hen in a battery cage" (Singer, 1994, p. 241). This might be possible to some extent, although a risk always exists of projecting human perspectives on animals and ending up in anthropomorphism. However, the ability to suffer is special. It is not just another characteristic along with the capacity for language or mathematics. Suffering is no doubt ethically relevant, and moral responsibility requires that we empathize with suffering animals.

On the other hand, it is important not to be naive regarding our ability to feel what laboratory animals feel (Nuffield Council on Bioethics, 2005, pp. 62–64). Different species may react differently to the same stimuli. Our moral imagination must be based on scientific evidence. We have reasons to believe that animals can suffer to the extent that they have a central nervous system and exhibit a behavior that, in human beings, is associated with pain. This view is further supported from an evolutionary perspective. Due to evolution, great genetic similarities exist between different species.

It is vital to consider not only the subjective feelings but also the biological functioning and natural living of the animals (see Chapter Five). Whether an animal exhibits weight loss or a reduced ability to move as a result of an experiment may be ethically relevant, even if the animal does not feel pain. Moral imagination may also require that we consider aspects other than animal welfare, such as the integrity of animals.

Moreover, moral imagination may require that we recognize a plurality of values regarding animals. Examples are animal welfare, animal integrity, and interspecies justice. Regarding animal welfare it is important to note that this may not only be understood as a value but also as a scientific concept. I will discuss the relation of science and ethics with regard to this concept in Chapter Five.

Finally, a key idea of imaginative casuistry is case-by-case balancing. This makes this approach especially appropriate in the ethics of animal experimentation. In animal experimentation the details of individual experiments may be of central importance. For example, the precise purpose of an experiment and the particular animal species to be used has to be taken into consideration. Moral imagination is pivotal in exploring different aspects of an animal experiment and in carrying out case-specific ethical balancing.

Four

THE SCIENTIFIC VALUE OF ANIMAL EXPERIMENTATION

A key issue in the ethics of animal experimentation concerns the scientific value of such experimentation. To what extent can results from animal experiments be extrapolated to human beings? If it is impossible to extrapolate the results to human beings, all ethical arguments in favor of animal experimentation for human benefit would be undermined. Animal experimentation aiming at human benefit would not be ethically acceptable. It would make ethical arguments against animal experimentation for human benefit redundant. But if extrapolation to human beings is possible—at least to some extent—then the issue changes its character. Each animal experiment has to be assessed on its own merit, scientifically with regard to the reliability of extrapolation to human beings and ethically with regard to expected human benefit and animal harm.

1. Animal Experimentation in Present-Day Basic and Applied Research

Animal experimentation is at present a central part of basic and applied research. No sharp distinction can be made between basic and applied research, but we find prototypical instances of each. Basic research aims at increasing knowledge with no practical application in mind. Applied research on the other hand aims at the solution of practical problems. An example of basic research is research aiming at understanding the fundamental function of the cell. A prototypical example of applied research is testing of the safety and efficacy of a pharmaceutical drug under development. Basic and applied research are often closely related, however. Basic research may lead to applied research and applied research has given rise to basic knowledge.

In order to give a picture of the use of animals in present-day basic and applied research, let me present some statistics.

In the European Union, reports regarding the number of animals used for experimental and other scientific purposes in the Member States are published every five years. The most recent report, already referred to in the Introduction, was published in 2005 and concerns data from 2002 (note that France submitted statistics from 2001). The total number of animals used was 10.7 million. Mice were 51%, 22% were rats, and 3% consisted of other rodents. Other mammals amounted to 4%, among them primates to 0.1%. The proportion of fish was 15% and birds 6%. These animals were used for different purposes: 35% were used in fundamental biology studies, 31% in research and development in human medicine, veterinary medicine, and dentistry, 14% for production and quality control in human medicine and dentistry, and 10% in toxicological and other safety evaluation (Commission of the European Communities, 2005).

What characterizes present-day research on animals? Let me give a brief overview. In this overview I draw heavily on the excellent presentation by the Nuffield Council on Bioethics in its report *The Ethics of Research Involving Animals* from 2005 (pp. 83–184).

A. Basic Research

Basic research includes a wide range of studies, such as behavioral studies, physiological studies, developmental studies, genetic studies, and the development of research tools (this section is based on information provided in Nuffield Council on Bioethics, 2005, pp. 87–103, 171–172). Some of these studies are merely observational, others highly invasive. Some research is carried out primarily to increase our knowledge about animals and their behavior. Other types of research aim at increasing our understanding of basic biological processes. To some extent this basic knowledge may lead to applications that are of direct human benefit.

Observational studies on animals in their natural habitat are conducted in order to understand, for example, social interactions between animals. Behavioral studies are also carried out in laboratories. In this research for example mazes are used to investigate rodent learning and memory.

Physiological studies may involve surgery or drug treatment. The aim is to understand bodily function at the physiological, cellular, or molecular levels. The investigations provide knowledge about the endocrine system, the immune system, and the nervous system. This knowledge may also have important applications for human benefit. For example, studies of graft rejection in immunodeficient rodents have been vital in the development of human organ transplantation.

Animal experimentation has also contributed to our knowledge of the human nervous system. For example, monkeys have been used in studies of how activity in groups of brain cells in the motor cortex control hand and finger movements. The aim has been to understand how stroke can impair use of the human hand. This kind of research has led to the development of treatment to reduce the symptoms of Parkinson's disease (Kumar *et al.*, 1998; Rodriguez-Oroz *et al.*, 2004).

The study of animal development has contributed to our understanding of human development. Embryos from chicken, zebra fish, rodent, and frog are used to obtain knowledge about the function of genes in developmental processes. Genetically modified mammalian embryos have also been generated for this purpose. Developmental studies have also been undertaken on young and adult animals, in particular in mammals, where major development takes place after birth.

Genetic studies are an increasingly important part of animal research. Ever more genetically modified animals are produced in order to obtain knowledge about gene function. As I mentioned in the Introduction, animal cloning may also be useful. Examples of potential uses are providing organs for xenotransplantation, pharming, production of "copies" of farm or sport animals with special traits, and replacement of deceased companion animals.

Finally, animals are used in the production of research tools such as antibodies, which can be used to identify, quantify or purify a substance. In the production of antibodies against a particular antigen, the animal is repeatedly immunized with the antigen in combination with an adjuvant, that is, an immunostimulant. The antibodies are then harvested from the blood.

Basic research may sometimes lead to unintended and unexpected medical applications of great human benefit. One example given in the Nuffield report concerns narcolepsy, the cause and nature of which were unknown until recently. Two research groups discovered independently a neurotransmitter produced by the hypothalamus. Neither of the groups was working on narcolepsy. When the gene for the neurotransmitter was knocked-out in mice, the mice developed narcolepsy (Sakurai *et al.*, 1998; De Lecea *et al.*, 1998). Later another group that studied narcolepsy in dogs discovered a gene coding for a membrane receptor for one of two forms of the neurotransmitter (Lin *et al.*, 1999). On the basis of these findings about narcolepsy in mice and dogs, two other groups analyzed the brains of deceased human beings who had suffered from narcolepsy. They found that the cells in the hypothalamus producing the neurotransmitter were decreased or even absent (Peyron *et al.*, 2000; Thannickal *et al.*, 2000). At present it is believed that narcolepsy in human beings is caused by autoimmune destruction of these cells.

B. Animals as Disease Models

Animals are used in the study of human diseases and their causes, and to develop therapies (this section is based on information provided in Nuffield Council on Bioethics, 2005, pp. 105–129, 173–174). *In vitro* methods are also used, but many scientists argue that entire animals are necessary in understanding the complex and dynamic interactions between molecular, cellular, and organ systems. The animals used as disease models can be obtained by discovery of spontaneous mutations, selective breeding or genetic modification.

An example is animal models for rheumatoid arthritis, one of the most common autoimmune diseases. Rodent models with induced arthritis have contributed to the discovery that an immune molecule called TNF plays a central role in the inflammatory process. The animals experienced a painful swelling of the paws and damage to the cartilage. Several methods were tested on the models. The goal was to neutralize the inflammatory reactions by blocking TNF through administration of antibodies. This strategy reduced the inflammation and damage. Clinical trials were undertaken in human beings and many people have been treated effectively with this antibody therapy (see Vilcek and Feldmann, 2004).

Genetically modified animals are increasingly being used as disease models. Examples are models for diabetes, deafness, psychiatric disorders, neurodegenerative disorders, and cancers.

In the study of genetic diseases the mouse is especially useful, because it has strong genetic similarities with human beings: 99% of genes in mice have direct counterparts in human beings. Other species with suitable genomes for comparative studies are the zebra fish and the rat.

Mouse models have made it possible for scientists to study the relationship between mutations and the nature and severity of the disease they cause. One example is the glucokinase gene in diabetes (Toye *et al.*, 2004). Another example is the mouse model shaker-1 that led to the discovery of a gene causing hearing loss in both mice and human beings (Gibson *et al.*, 1995). Mouse models are also of vital importance for investigating how a disease can produce varying symptoms in different individuals. Indirect changes, for example in protein or hormone levels, may be more suitable therapeutic targets than the genes themselves. This has proved to be the case in patients with neurodegenerative disorders.

C. Animal Use by the Pharmaceutical Industry

Animal models play an important role in research and development of new drugs (this section is based on information provided in Nuffield Council on Bioethics, 2005, pp. 131–151, 175–176). Relatively small numbers of animals are used in the early stages of drug development. Many of them are genetically modified mice. They are used to investigate whether, for example, specific receptors might respond to chemical substances that can be developed into new medicines. Animal models can also be used to test how human beings affected by a disorder react to different chemical compounds.

Relatively high numbers of animals are used in the process of characterizing promising candidate medicines. Before a potential drug is tested in clinical trials, it must be ensured that it exhibits an acceptable balance of safety and efficacy. This usually requires data from animal tests. Even when a medicine is in clinical trial, animal tests may continue to be undertaken. For some compounds such as vaccines, animal testing is required for each batch that is produced, to ensure safety and efficacy.

Research and development of a new pharmaceutical drug takes commonly about 10–15 years. Let us take a closer look at the process.

In the first stage of the process, useful targets are identified. These can be disease-related genes or proteins that function as receptors for active molecules of new medicines. In the second stage, possible medicines are identified. In both stages findings from basic research are applied. Of central importance are new automated technologies. Compounds that might interact with the targets are subjected to high-throughput screening, that is, automated testing. Starting from hundreds of thousands of compounds, the end-result of the screening may be around 1,000 compounds. The molecules are screened against animals, animal tissues, and cloned human receptors. Only a few animals are used. Animal tissues are used for some tests. Cloned human receptors are preferred.

In the next two stages, the pharmacological properties of potential medicines are investigated. The third stage is characterized by "hit-to-lead chemistry," that is, the potential medicines ("hits") are converted into a few promising compounds called "leads." In the fourth stage, the leads are optimized by synthetic chemical modification. These refined leads may be used clinically. The majority of the animals used by the pharmaceutical industry are involved in these two stages. Genetically modified animals are increasingly common.

The fifth stage aims at determining whether promising compounds can be tested in clinical trials on human beings. The safety of the candidate medicines is ensured. This toxicity-testing completes the pre-clinical phase of the development process. The potential medicine shows an acceptable balance of safety and efficacy.

The following stages consist of clinical studies of human beings (phases I–IV). In phase I studies, the safety and dosage of the potential medicine are tested on 20–100 healthy volunteers. In this way it is found out how well the active compound is tolerated in human beings. Phase II involves trials on 100–500 patients regarding safety and efficacy. Phase III trials involve 1,000 to 5,000 patients and aim at establishing safety and efficacy more accurately. Throughout all three phases safety tests on animals continue to be undertaken. The data from pre-clinical and clinical studies are then submitted to regulatory agencies for approval. When a medicine is authorized, the marketing stage starts, but phase IV clinical trials are conducted in order to monitor long-term effects in large numbers of patients.

D. Animals in Toxicity Testing

Animals, mainly rats and mice—but also non-rodent species—are used in safety assessment of substances such as pharmaceutical drugs and chemicals used in households, agriculture, and industry (this section is based on information provided in Nuffield Council on Bioethics, 2005, pp. 153–167, 176). The chemicals are evaluated for their potential to cause irritation, physiological reactions, cancers, developmental effects on fetuses, and effects on fertility. Specified doses are given to animals, and the results are then extrapolated to human beings. Some tests concern single high doses, others long-term exposure. The tests may lead to restrictions regarding how the drugs and chemicals may be used.

Many different types of toxicity tests involving animals exist. In acute toxicity studies, adverse effects that may occur on first exposure to a single dose of a compound are examined. Some tests concern effects of contact with the skin or eye. Others concern the effects on organs of a substance that is swallowed, inhaled, injected, or absorbed through the skin. Several alternative methods have been developed in order to reduce the number of animals used. In some cases, signs of significant toxicity have replaced death as the endpoint.

Repeated-dose toxicity studies are also carried out, generally on mice and rats. The tests are conducted for different periods of time, commonly 28 days but also 90 days to one year. The results indicate the highest dose without significant adverse effects.

Other tests concern carcinogenicity. In these tests rats and mice are exposed for up to two years and the incidence of tumors is evaluated. In this way, the risk for cancer in human beings can be estimated.

Testing for genotoxicity, that is, the potential of a substance to interact with the genome, causing cancer or heritable mutations, is commonly carried out *in vitro* on bacteria or mammalian cells. Animals are tested only when these *in vitro* tests have given positive results.

Effects on reproduction and development are also subjected to testing. In two-generation reproduction studies, rats are given repeated oral doses throughout sexual maturation into adulthood. The rats mate, and the females are dosed until the pups are weaned. This procedure is then repeated with the pups, and the second generation is assessed. In developmental toxicity studies, the effects on the unborn of exposure of the mother to a substance during pregnancy is tested. Here, rabbits are used in addition to rats because rodents do not respond, or respond variably, to human teratogens such as thalidomide (for further discussion of this example, see below).

In animal toxicity testing, several sources of uncertainty exist. Species, strain, and gender variations may make it difficult to transfer the results to human beings. The same holds true for scaling from small animals with a short lifespan that are tested with large doses to large human beings with a long lifespan obtaining small doses. The population of animals used in testing is commonly homogeneous, while significant variation often exists among human beings, affecting, for example, drug metabolism. Scientists need to take these sources of uncertainty seriously. If they do, animals can be useful models for predicting toxicity in human beings, or at least so it is argued (Nuffield Council on Bioethics, 2005, p. 158). This brings us to the controversial issue of the scientific value of animal experimentation.

2. Prototypical Cases of Scientifically Valuable Animal Experiments

When I talk about the "scientific value" of animal experimentation I mean instrumental value. The question is whether animal experiments have value as instruments for obtaining knowledge that can be of value for human beings, for example in the development of medical treatments. In addition, animal experimentation may be of value for animals, for example farm animals, although this will not be our focus here.

In relation to human beings, it is common among scientists to talk about "animal models." "Animal models" can be defined as animal systems that are believed to resemble human systems and are used in experiments to provide knowledge or hypotheses about human systems. The key condition for animal models is relevance. Relevance is a matter of similarity. The result of a particular animal experiment must be similar to the result of a corresponding human experiment. More precisely, the results must be *sufficiently similar*, but—as we will see below—what is considered to be sufficient similarity is a matter of controversy. In this context, Hugh LaFollette and Niall Shanks have made a distinction between two types of problems of relevance that may arise, an ontological problem and an epistemological problem. The ontological problem is whether human beings and animals are sufficiently similar. The epistemological problem is how to determine that they are sufficiently similar (LaFollette and Shanks, 1996, p. 22). Animal models need to be relevant in both respects.

A common way of showing the scientific value of animal experimentation is to give some successful examples. We have already seen some examples in the overview of present-day animal experimentation above. Let me give a few more, namely some of those provided by The Boyd Group (Smith and Boyd, 1991). In its argument in favor of the scientific value of animal experimentation, this group refers to a "glossary" provided in a report from the American Medical Association's Council of Scientific Affairs from 1989. They state:

> This glossary covers medical advances through research on aging (including improved understanding of the pathology of Alzheimer's disease), AIDS, anaesthesia, autoimmune diseases, basic genetics, behaviour (including the development of neurosurgical procedures), diseases of and defects in the cardiovascular system, childhood diseases, cholera, convulsive disorders, diabetes, gastrointestinal tract surgery, hearing, haemophilia, hepatitis, infection, malaria, muscular dystrophy, nutrition, ophtalmology, organ transplantation, Parkinson's disease, treatment of pulmonary disease and injury, prevention of rabies, radiobiology, reproductive biology (including the development of the contraceptive pill), skeletal system (including orthopaedic surgery), treatment of spinal cord injuries, toxoplasmosis, trauma and shock, yellow fever, and virology.
>
> Whether the benefits alluded to above could have been achieved without the use of animals seems unlikely. This, however, is the kind of hypothetical historical question which no one is in a position to answer (Smith and Boyd, 1991, p. 27).

The problem is whether such historical examples of successful animal experiments truly justify present-day experiments. Smith and Boyd stress:

> The undoubted benefits of animal use in the past, however, do not mean that the continued and unquestioning use of animals in biomedical research today is thereby also morally justified (Smith and Boyd, 1991, p. 27).

This is an important clarification. Historical examples of successful animal experiments may indicate that it is wrong to maintain that all animal experimentation is useless, but present-day animal experiments must be judged on their own merit.

In his argument for the strong human priority prototype, Cohen therefore goes a step further by providing reports of "exciting medical and biological investigations that are under way *now*, in the effort to cope with human disease" (Cohen in Cohen and Regan, 2001, p. 86). He gives examples of promising up-to-date animal experiments concerning heart failure, vaccines, Alzheimer's, Parkinson's, ALS and other neurodegenerative diseases, obesity, sleep, organ transplants, diabetes, genetic diseases, cancer, and emphysema (Cohen in Cohen and Regan, 2001, p. 86–117).

Let us turn to arguments against the scientific value of animal experimentation.

3. The *Con* Argument from Causal Disanalogy

Several arguments against the scientific value of animal experimentation deny the possibility of extrapolating data from animal experimentation to human beings. C. Ray Greek and Jean Swingle Greek point out:

> Data obtained from animal models in biomedical research, for the purpose of evaluating the safety and effectiveness of pharmaceutical drugs, testing carcinogens, conducting research on human diseases such as AIDS, and so forth, cannot be reliably extrapolated to humans (Greek and Greek, 2002, p. 25).

Their examples indicate that "data" is understood in a broad sense, and they argue that data in this broad sense obtained from animal models cannot be "reliably" extrapolated to human beings.

Another argument focuses on causal mechanisms instead of data in a broad sense. It stresses causal disanalogy, that is, that the causal mechanisms may not be similar in animal models and in human beings. The philosophically most important version of this argument has been developed by Hugh LaFollette and Niall Shanks in their book *Brute Science: Dilemmas of Animal Experimentation* (1996). They summarize their view as follows:

> The presence of causal disanalogies undermines the claim that animal research is of immediate and direct relevance to human biomedical phenomena. More specifically, these disanalogies will undercut claims about the direct benefits of applied research—like predictive toxicology and teratology—which aims to make *predictions* about human biomedical phenomena (LaFollette and Shanks, 1996, p 107).

Let us take a closer look at their argument.

A. CAMs and HAMs

The central concept in the argument is that of causal analog models (CAMs). In order to make possible predictions about human beings, animal models need to be strong causal models, that is, there has to be strong causal analogy or, in others words, completely similar causal mechanisms. There can be no causal disanalogies.

They articulate the following schema for causal analogical arguments:

> X (the model) is similar to Y (the subject being modeled) with respect to properties [a, ..., e]. X has additional property f. While f has not yet been observed directly in Y, it is likely that Y also has the property f (LaFollette and Shanks, 1996, p. 113).

According to LaFollette and Shanks, CAMs must satisfy the following conditions:

> (1) the common properties [a, ..., e] must be properties which (2) are causally connected with the property... f... we wish to project—specifically, ... f... should stand as the cause(s) or effect(s) of the features ... [a, ..., e] in the model (LaFollette and Shanks, 1996, p. 112).

In addition, they state also a third condition:

> (3) there must be no causally relevant disanalogies between the model and the thing being modeled (LaFollette and Shanks, 1996, p. 113).

This third condition is of key importance to their argument. They question whether animal models ever satisfy this condition. Their conclusion is articulated like this:

> We have seen that there are good reasons to think animal CAMs of human biomedical phenomena typically do not satisfy condition (3), and, thus, are not strong models (LaFollette and Shanks, 1996, p. 137).

Thus, LaFollette and Shanks seem to maintain that animal models are rarely if ever strong causal models. This problem is an ontological problem of relevance (*cf.* above).

They also point out an epistemological problem of relevance. They state that we can never know in advance whether condition (3) is fulfilled. They argue that the appropriateness of a model organism for extrapolation to human beings cannot be established without already knowing what we hope to learn from the extrapolation (LaFollette and Shanks, 1996, p. 23). This is sometimes called "the extrapolator's circle" (Steel, 2008, p. 4). In a sense, this is the crucial argument. It concerns what we can know. Even if (3) is satisfied we cannot know it in advance. We cannot even assume that it is satisfied (LaFollette and Shanks, 1996, p. 118).

Their arguments against animal models ever being strong causal models are of two types—the theory of evolution and empirical evidence—and will be discussed below.

Strong causal models are not the only option. Another option is weak causal models. In this case, the causal mechanisms are not completely similar. Causal analogies and disanalogies exist. This is how many scientists understand animal models. Differences are acknowledged but it is argued that they may be reduced by means of scaling. Scaling is adjusting for purely quantitative differences between species, for example, differences in body weight and metabolic rates. Scaling is supplemented by the idea that some animal species are more appropriate as models for a particular problem than others. The ambition is to find an animal species that is the best for solving a particular scientific problem.

LaFollette and Shanks accept that animal models can be weak models, that is, that there can be both analogies and disanalogies between animal models and human beings. But they deny that this is sufficient for making reliable predictions in applied research such as toxicological and teratological research. Weak models cannot serve the scientists' experimental purposes, if these purposes are to make predictions about what will happen in human beings based on findings in animals (LaFollette and Shanks, 1996, pp. 65–67, 138, 142–150).

The argument of LaFollette and Shanks against the idea that animal models as weak causal models can be used to predict what will happen in human beings is based on dynamic systems theory. I will discuss this argument below.

In sum, LaFollette and Shanks argue that in order to allow reliable predictions animal models must be strong models, because there can be no causal disanalogies. But we have good reasons to think that animal models rarely if ever can be strong models. Moreover, weak models are not sufficient, because if some causal disanalogies exist, we cannot make reliable predictions.

LaFollette and Shanks also present a third option, namely to view animal experimentation as a way of generating hypotheses about human biological phenomena. LaFollette and Shanks call such animal models hypothetical analog models (HAMs), and maintain that such models may have an important function in basic research. For example, animal models were useful in early studies of the DNA structure and in the development of immunological theory (LaFollette and Shanks, 1996, pp. 193–195).

The statement that animal models may be useful as generators of hypotheses about human beings—HAMs—appears quite uncontroversial. The claim that animal models cannot be CAMs is much more controversial and in need of further discussion. Let us therefore investigate their arguments in more detail. LaFollette and Shanks support their argument from causal disanalogy by citing empirical and theoretical evidence. The empirical part consists of examples from the history of animal experimentation. The theoretical support is gathered from different scientific theories such as the theory of evolution, dynamic systems theory, and genetics.

B. Empirical Support for Causal Disanalogy

As in the case for animal experimentation, a common way of showing that animal experimentation is misleading is to give examples. In their introductory discussion of "the problems of relevance," LaFollette and Shanks mention that morphine sedates human beings but stimulates cats, that penicillin has adverse effects on guinea pigs and hamsters, that benzene causes leukemia in human beings but not in mice, and that aspirin causes birth defects in rats and mice, poisons cats, but does not affect horses (LaFollette and Shanks, 1996, pp 25–26). They also mention the thalidomide disaster (LaFollette and Shanks, 1996, pp. 14–15, 26–27; see below). In giving further empirical support for their causal disanalogy argument, they cite scientists from different scientific fields such as toxicology, teratology, endocrinology, virology, and stroke research. Throughout the presentation they stress the "pervasiveness" of causal disanalogies between different species (LaFollette and Shanks, 1996, pp. 120–129).

LaFollette and Shanks conclude that this empirical evidence speaks against animal models being strong causal models. Animal models can only be weak causal models. But this empirical evidence also speaks against the possibility of making reliable extrapolations on the basis of such weak models.

C. Theoretical Support for Causal Disanalogy

We have seen examples both in favor of and against the scientific value of animal experimentation. LaFollette and Shanks make this comment on giving such examples:

> Proponents cite cases that purportedly show that animal research has enormous utility; opponents cite cases that purportedly show that animal research has misled us or has failed to contribute to human health and well-being... Although the "examples game," when played well, is rhetorically effective, it is not as simple an argumentative strategy as it appears, and as both sides assume it to be... We must have some theoretical framework from within which we interpret the experimental results as successful (LaFollette and Shanks, 1996, pp. 29–30).

In order to support their argument from causal disanalogy, LaFollette and Shanks develop a theoretical framework based on different scientific theories. One of them is the theory of evolution.

This is how LaFollette and Shanks introduce the theory of evolution:

> Evolution is currently the unifying theory in modern biology. However, the modern physiologists' paradigm fails to understand or appreciate the implications of biological evolution for the practice of physiology. That should not be surprising since the current physiological paradigm was established by Bernard who ... rejected evolution. The paradigm continues to be wedded to biological reductionism (LaFollette and Shanks, 1996, p. 68).

We see here that LaFollette and Shanks find a tension between established physiology and the theory of evolution. Physiologists do not take the implications of the theory of evolution seriously enough. While physiology is characterized by "reductionism," this is not the case with regard to the theory of evolution. We will return to the non-reductionism of the theory of evolution below.

LaFollette and Shanks admit that:

> At first glance it appears the theory of evolution would guarantee that there would be no relevant differences that would undermine our ability to extrapolate from one species to another, especially phylogenetically close species. After all, the theory of evolution suggests that there exist important biological similarities between members of distinct species (LaFollette and Shanks, 1996, p. 111).

They stress that the theory of evolution may also have us expect differences between species:

> The ability of differently evolved creatures to achieve common biological functions through different causal means is ubiquitous. For example, evolutionary theory leads us to expect that members of distantly related species may employ different mechanisms to achieve the common function of gas exchange with the environment. These differences are most apparent when we contrast fish with mammals. However, even two organisms with lungs (mammals and birds) may have substantially different underlying causal mechanisms for exchanging gases with the environment (LaFollette and Shanks, 1996, p. 100).

LaFollette and Shanks argue that the theory of evolution has four "consequences" that we have to take seriously. The first consequence is that "from similarity of biological function we cannot infer similarity of underlying causal mechanism," the second that "from differences in causal mechanisms we cannot infer differences in functional properties," and the third that "from similar causal mechanisms (and values of casually relevant parameters) we cannot infer similar functional properties." The final consequence is that "although we cannot infer similarity of causal properties from similarity of functional properties, we can infer differences in causal properties from differences in functional properties" (LaFollette and Shanks, 1996, p. 100). The general point is that an asymmetry exists between causal mechanism and biological function. According to LaFollette and Shanks, this asymmetry supports the *con* argument from causal disanalogy.

In sum, LaFollette and Shanks maintain that the theory of evolution—in addition to empirical evidence—provides a strong argument against animal models being strong causal models.

They continue by arguing also against the predictive value of animal models as weak models. In this regard, LaFollette and Shanks turn to dynamic systems theory (or complexity theory). The reason for this is their focus on emergent properties. This is how they present the importance of such properties for an adequate theoretical perspective on animal experimentation:

> This emergence of biomedically significant properties at higher levels of complexity is crucial for a proper scientific understanding of animal experimentation .... Physiologists acknowledge that species may appear different. However, in a manner reminiscent of Bernard, they claim that despite these seeming differences, species are fundamentally similar. In one sense, they are undoubtedly correct. Biochemical (and metabolic) evolution has been very conservative. We know that all life is based on DNA and RNA and that metabolic pathways are roughly similar across species ... yet physiologists tend to make too much of these similarities in basic substrate and structure. We should not conclude that these similarities imply more general similarities higher up the biological hierarchy. The organizational complexity of biological organisms makes this inference questionable (LaFollette and Shanks, 1996, pp. 91–92).

Dynamic systems theory provides a general theoretical framework for understanding the emergence of new properties at higher levels of complexity. In this way it may explain the asymmetry of causal mechanisms and function, which is the key presupposition in the argument from causal disanalogy. In particular this argument—according to LaFollette and Shanks—speaks against the predictive value of animal models used in applied research. Due to complexity and non-linearity, reliable predictions about human biomedical phenomena on the basis of animal models are impossible.

A theorist that probably more than any other has contributed to the theoretical understanding of hierarchical complexity in evolutionary biology is Stuart Kauffman (1993), and consequently he is among those that LaFollette and Shanks cite (LaFollette and Shanks, 1996, pp. 91–92). Kauffman tries to combine the theory of evolution with dynamic systems theory. Kauffman does so by supplementing natural selection as the key factor behind evolution with self-organization.

The third theoretical framework on the basis of which LaFollette and Shanks develop their argument from causal disanalogy is genetics. Their key focus is on gene regulation. Different types of genes exist. Structural genes code for proteins. Regulator genes turn structural genes on and off. LaFollette and Shanks stress that mutations in regulator genes may have radical effects. The turning on and off may be due to developmental factors and these are often different in animal models and human beings. In this way they may be the source of causal disanalogies (LaFollette and Shanks, 1996, p. 185). LaFollette and Shanks reject genetic determinism and emphasize the complex interaction with other genes and with the environment. They use this as an additional argument in their criticism of the view that transgenic animals may provide more reliable models than ordinary laboratory animals. The complex interaction may influence the expression of the transgene and make predictions impossible (LaFollette and Shanks, 1996, pp. 190–192).

4. Objections to the *Con* Argument from Causal Disanalogy

The argument from causal disanalogy should be taken very seriously and it is warranted to discuss some objections. However, I will not criticize particular historical examples. For instance, I am aware of the criticism against Greek and Greek that they misinterpret history (Festing, 2001; Guerrini, 2004). I regard this as a problem for historians of science and medicine. I will focus on scientific and philosophical arguments.

Let me also make it clear at the outset that these are objections that I direct against their argument. I believe that LaFollette and Shanks are too pessimistic regarding the possibility of extrapolating from animal models to human beings, but their contribution to the discussion on animal experimentation is important and should function as a reminder not to be naive regarding extrapolation.

A. Some Extrapolations Are More Justified than Others

An important objection to LaFollette and Shanks is directed toward their third condition for strong CAMs, that is, that no causally relevant disanalogies whatsoever exist between the model and the thing being modeled. The objection is that this ontological condition is too strict. It is rarely if ever satisfied in animal experimentation. Countless disanalogies exist between human beings and any animal species, some of which are causally relevant. This condition would make illegitimate extrapolation not only from animals to human beings but also from human beings to other human beings. Even between individual human beings countless disanalogies exist. This fact is the motivating reason behind the emerging field of pharmacogenomics, which studies the genetic differences among individuals that produce different responses to drugs. Moreover, causally relevant differences exist even within a single organism at different stages of life. Not even extrapolations from past to future in the life of a single person would be justified (Steel, 2008, p. 93). By their extremely strict ontological requirement LaFollette and Shanks create an ontological problem that is more difficult to overcome than it need be. We need a different conceptual framework (see below).

LaFollette and Shanks accept animal models only for generating hypotheses in basic research (HAMs), but to generalize and state that all uses of animal models in applied research lack scientific value is—in my opinion—to go too far. We should search for other ways of conceptualizing animal models in applied research.

Let us first look at the empirical aspects. To what extent have animal models proven effective in applied research? Zambrowicz and Sands carried out an interesting retrospective study. They focused on the use of knock-out mice in drug development. They found that the 100 best-selling drugs had 43 human biochemical targets and that the genes for 34 of these targets had been knocked out in mice. Of these 34 knock-out models, 29 (85%) provided a direct correlation with the therapeutic effect in human beings. The remaining cases were not useful models, since they exhibited early lethality or unrelated abnormalities (Zambrowicz and Sands, 2003; also referred to in Nuffield Council on Bioethics, 2005, p. 182). An objection is that the genetically modified animals were produced after the drugs were developed and the causal mechanisms were already known, but Zambrowics and Sands argue that at present several "prospective" uses of knock-out mice are promising (Zambrowics and Sands, 2003).

A good starting-point for a search for an alternative conceptual framework would be to recognize that probably no such thing exists as "the perfect animal model" but that animal models can be better or worse. As stated in the Nuffield report:

> An animal need not share all properties of humans to be an effective model. It is sufficient for the model to be similar in relevant aspects of the disease being studied (Nuffield Council on Bioethics, 2005, p. 138).

If animals are to be useful as models, it is only necessary that some relevant aspects of their biological processes are similar to those of human beings. Weak models can be used to make some predictions about human biological phenomena. Weak models can be more or less reliable. LaFollette and Shanks neglect to recognize that within the category of weak models there may be relevant differences.

In order to determine which animal models are better, we must focus on causal mechanisms instead of the observed effects as such. Here, Daniel Steel suggests a method that he calls "comparative process tracing." According to this method, providing evidence for the suitability of an animal model requires comparisons only at stages (phases) in the causal mechanism in which differences are likely to occur. The greater similarities at these stages, the stronger are the bases for extrapolation of effects (Steel, 2008, pp. 78–100).

A prototypical case is animal experimentation pertinent to carcinogenic effects of aflatoxin B1 in human beings (liver cancer). Here comparative process tracing indicates that the rat is a better model than the mouse. First, we should distinguish two stages in the metabolic mechanism. In phase I, a compound is changed chemically in a way that makes it more polarized and consequently more easily excreted. In phase II, the modified compound is conjoined with a macromolecule that makes it less toxic and even more easily removed (Steel, 2008, pp. 88–89). Phase I metabolism of aflatoxin B1 is similar in all three—human, rat, mouse—but phase II metabolism is much more similar in rat and human beings. This means that what is carcinogenic for the rat is probably carcinogenic for human beings, too. However, the quantity of DNA adducts resulting from aflatoxin B1 is less in rat than in human beings, and this difference suggests that the carcinogenic effects on rat are less than in human beings. Thus, we have good reason to believe that aflatoxin B1 is carcinogenic in human beings, but we cannot predict exactly how much. This shows that a model might be a good basis for qualitative extrapolations but not for quantitative ones (Steel, 2008, pp. 91–94).

This illustrates how the epistemological problem of the extrapolator's circle can be avoided. In order to establish the suitability of the animal model, we do not need to know everything about the causal mechanism in human beings, in which case the extrapolation would be unnecessary. Only a few key features of the mechanism in human beings need to be examined, and this would fall far short of what we can hope to learn from the extrapolation (Steel, 2008, pp. 94–96).

In practice scientists seldom start from scratch. They already have some experience from the particular field of interest and know which animal species are useful in that field. They can also easily find information about how causally relevant mechanisms differ between human beings and different animal models, and regarding which types of compounds (Steel, 2008, p. 88).

In conclusion, Steel's alternative conceptual framework of comparative process tracing appears promising. It captures much better the actual role of animal models in applied research.

Let me finally comment on the thalidomide disaster. This disaster is often referred to by those criticizing the scientific value of animal experimentation (for example, LaFollette and Shanks, 1996, pp. 14–15, 26–27; Greek and Greek, 2000, pp. 44–47). There appears to be a lot of misunderstanding regarding thalidomide and animal experimentation. Two points are important to note. First, thalidomide was never tested on pregnant animals before it was used on human beings (this was not a legal requirement at the time). Moreover, after the drug had been withdrawn because of its disastrous effects on human beings, these effects were also discovered in, for example, mouse, rat, hamster, rabbit, rhesus monkey, and baboon (Nuffield Council on Bioethics, 2005, p. 145). My conclusion is that the problem was not animal experimentation but instead that there was too little animal experimentation. If thalidomide had been tested on pregnant animals, the disaster might have been avoided. The thalidomide case is actually an argument in favor of animal experimentation rather than against.

The argument of LaFollette and Shanks is not a tenable argument for the conclusion that extrapolation from animal models to human biological phenomena is rarely if ever justified. The "extrapolator's circle" can sometimes be avoided. Some extrapolations are more justified than others. Sometimes sufficient causal analogy exists, sometimes not. Sometimes an extrapolation from animals to human beings is justified, sometimes not. For this reason, we must always focus on particular cases.

B. A Combination of Methods May Strengthen Extrapolation

Another objection is that a combination of different methods may allow more accurate extrapolation to human beings. In applied research, for example pharmacological research on the physiological effects of drugs, we should not use a single method—animal or non-animal—but several alternative methods such as investigations of whole animals, isolated organs, tissue culture, biochemical analysis, and mathematical modeling. Any single model may be weak, but a combination of models may strengthen extrapolation.

C. Alternatives May Be Less Reliable

A third objection may be formulated in a series of questions: Maybe the alternatives are less reliable? Maybe using isolated organs, tissue culture, biochemical analysis, or mathematical models is not more reliable than using a whole animal? To what extent are the alternatives causal analog models in the strong sense? What is their relative weakness with regard to predictive power? Alternative methods may be important but they may have their particular limitations. Even human beings are not necessarily strong models for other human beings. Individual drug responses may vary substantially. A drug dose that is harmless to one person may be lethal to another. This means that we must weigh risks and benefits of the alternatives and compare them with animal experimentation.

D. Criticism of the Theory-Driven Approach

LaFollette and Shanks stress the importance of a theoretical framework in order to assess examples of reliable and unreliable animal experiments (LaFollette and Shanks, 1996, pp. 30, 265, 269). An objection is that they overstate the importance of theory. Theory cannot determine which kinds of animal experimentation work and which do not work. The theoretical framework of LaFollette and Shanks, which is a mix of the theory of evolution, dynamic systems theory, and genetics, does not provide a sufficient basis for their claim that animal experimentation only works as a method for generating hypotheses in basic research and not for testing hypotheses in applied research. Even if the theory of evolution highlights not only similarities but also differences among species, even if dynamic systems theory stresses complexity and non-linearity, and even if complex interaction occurs between genes, and between genes and the environment (cellular and extra-cellular), some animal models may still sometimes say something that is relevant for human biological phenomena. Some extrapolations might still be possible. Theories are too abstract, general, and indeterminate. From these theories it is not possible to draw any conclusions about the reliability of animal experimentation. This is an empirical and practical matter. The theory-driven approach of LaFollette and Shanks is not in line with the pragmatic and tentative character of empirical science. In assessing animal experimentation, a pragmatic approach appears more appropriate than a theoretical one. To some extent animal experimentation works, and this is what is important.

In addition, let me also object to the way LaFollette and Shanks portray the relation of the theory of evolution and dynamic systems theory. In their presentation they mix quotations from Kauffman and other proponents of a more non-reductionistic theory of evolution such as Ernst Mayr and Stephen Jay Gould with quotations from proponents of the traditional, more reductionistic neo-Darwinian theory such as John Maynard Smith and Richard Dawkins. This is misleading. It gives the reader the mistaken impression that the dynamic systems approach with its focus on complexity and self-organization is an established part of the present-day theory of evolution. As I have shown elsewhere, two competing versions of the theory of evolution can be found in modern biology: "natural selection models" and "natural selection plus self-organization models." The latter version is controversial and does not represent the mainstream (Nordgren, 1994). Let me make it clear that I support the combined type of model. This means that I also come close to the version of LaFollette and Shanks. However, I do not believe that this theoretical approach can settle the issue of which kinds of animal experimentation are scientifically appropriate. This is a pragmatic issue.

Let me also comment on the argument from genetics. A great deal has happened in this field since LaFollette and Shanks published their book in 1996. The human genome has been sequenced (International Human Genome Sequencing Consortium, 2001; Venter *et al.*, 2001) and also the genomes of many other species including the mouse (Mouse Genome Sequencing Consortium, 2002), the rat (Rat Genome Sequencing Project Consortium, 2004), and the chimpanzee (The Chimpanzee Sequencing and Analysis Consortium, 2005). The main focus is now on functional genomics and comparative genomics, and it is becoming increasingly clear why there often are problems in extrapolating from animals to human beings because of gene regulation and epigenetics. So, in a sense these new results strengthen the argument of LaFollette and Shanks. We should not overstate their importance, however. As I have argued, some extrapolations are more justified than others. Some animal species are more similar to us also regarding gene regulation than others and can be used as models. However, gene regulation and epigenetics may also create problems in genetic engineering of animals, leading to unintended results and thereby to unintentional animal suffering. I will discuss these animal welfare problems in Chapter Six.

E. Conclusion

These objections show that the argument of LaFollette and Shanks is untenable. LaFollette and Shanks misunderstand the significance of causal disanalogy. Causally relevant differences always exist between human beings and animal models, but this does not make all extrapolations impossible. A more well-founded conclusion is that some extrapolations from animal models to human beings are more justified than others. But this is a pragmatic issue instead of one that can be solved on the basis of general theoretical considerations, and it has to be determined from case to case by means of the comparative process tracing explicated by Steel.

5. The 3Rs: Replacement, Reduction, Refinement

If we decide to carry out animal experiments, how should these experiments be designed in order to take animal welfare seriously? Several aspects should be considered. Let us take a look at the so-called 3Rs.

In 1959 W.H.S. Russell and R.L. Burch published their extremely influential book *The Principles of Humane Experimental Technique*. In this book, they presented their famous "3R approach," focusing on "Replacement," "Reduction," and "Refinement." The Universities' Federation for Animal Welfare (UFAW) considers this book so important that they have published a special edition (Russell and Burch, 1992).

Russell and Burch define the 3Rs as follows:

> Replacement means the substitution for conscious living higher animals of insentient material. Reduction means reduction in the numbers of animals used to obtain information of a given amount and precision. Refinement means any decrease in the incidence or severity of inhumane procedures applied to those animals which still have to be used (Russell and Burch, 1992, p. 64).

The principle of replacement implies that scientists should try to replace sentient animals with, for example, non-sentient animals, computer models or in vitro cultures. The principle of reduction states that no more animals should be used than necessary. Refinement involves taking the welfare of the laboratory animals seriously and trying to reduce their pain and distress.

In one sense, the 3Rs have a logical order. First, we should try to avoid using animals by using alternative methods (replacement). For those research purposes for which this is not possible, we should reduce the number of animals that are to be used (reduction). For those animals that are used, we should minimize the pain and distress (refinement).

The 3Rs are also interrelated in such a way that adjusting one can affect the others (*cf*. Nuffield Council on Bioethics, 2005, p. 189). For example, minimizing the pain and distress of animals may make it possible to use fewer animals.

An indication of the significance of the 3R approach is the fact that it has become an important objective of the European Union regulations. The Council Directive "on the approximation of laws, regulations, and administrative provisions of the Member States regarding the protection of animals used for experimental and other scientific purposes" states that

> experiments may not be performed if another scientifically satisfactory method of obtaining the result sought, not entailing the use of an animal is reasonably and practically available. The choice of species shall be carefully considered. In a choice between experiments, those which use the minimum number of animals, involve animals with the lowest degree of neuro-physiological sensitivity, cause the least pain, suffering, distress, or lasting harm and which are most likely to provide satisfactory results shall be selected. All experiments shall be designed to avoid distress and unnecessary pain and suffering to the experimental animals (Council Directive 86/609/EEC).

In this quotation, all 3Rs can be found. First, if possible, methods without animals should be used (replacement). Second, the number of animals should be minimized (reduction). Third, unnecessary suffering should be avoided (refinement).

6. Implications of the Five Prototypes for the 3Rs

What are the implications of the five ethical prototypes of animal experimentation for the 3Rs? At first glance, the human dominion prototype may seem to imply a rejection of the 3R approach. Carruthers (1992) views—at least tentatively—animal pain as non-conscious. However, respect for animal lovers may make him support this approach, although scientists have no direct duties to animals.

The strong human priority prototype accepts that we should try to minimize animal suffering in experimentation, but does not accept that the concern for animal welfare leads to not conducting certain animal experiments for human benefit. Cohen is explicitly critical of the principles of replacement and reduction (Cohen, 1994). He is in favor of more animal experiments in order to help people with diseases.

The weak human priority prototype is more in line with the 3R approach, accepting that this approach may not only influence how animal experiments are carried out but also whether they are to be carried out. Some experiments are not acceptable, because they cause too much animal suffering.

Both the prototype of equal consideration of interests and the animal rights prototype maintain that the 3Rs are good, but stress that they are insufficient. Regan states explicitly that "it is not enough first conscientiously to look for non-animal alternatives and then, having failed to find any, to resort to using animals. Though this approach is laudable as far as it goes, and though taking it would mark significant progress, it does not go far enough" (Regan, 1983, p. 385).

7. The 3Rs: Practical Implications

The arguments in favor of the 3R approach are of two types. Scientific arguments stress that there should be no waste of animals and that the effectiveness should be optimized in order to achieve scientific goals. Ethical arguments emphasize that animals should be treated in an ethically acceptable manner. The scientific and ethical arguments go hand in hand. Let us investigate each of the 3Rs from a scientific and ethical point of view.

A. Replacement

There can be different kinds of replacement. A distinction can be made between complete and incomplete replacements.

In complete replacements, the alternative methods do not involve any use of animals or animal-derived biological material whatsoever. Examples are mathematical and computer models. These may be used to model the biological activity of not only substances but also biological systems and processes. Other examples are studies involving human beings such as epidemiological studies, research on individual human subjects, and *in vitro* studies on human cells, tissues, and organs (Nuffield Council on Bioethics, 2005, p. 191). Recently, human stem cells have emerged as an important alternative. Embryonic and fetal forms of animals may also belong to the category of complete replacements. These are animals in a biological sense but not in the technical sense of the European Union regulation (see Chapter One). For the same reason, embryonic stem cells derived from animal embryos may also be included in this category. Similarly, the use of non-sentient invertebrates would be a complete replacement.

In incomplete replacements, biological material from animals is used, either from living animals or from humanely killed animals. This means that animals as such are not replaced but that with these methods experiments on living animals are replaced. Examples are animal cells, tissues, and organs (Nuffield Council on Bioethics, 2005, p. 191).

Within the category of incomplete replacements, I would like to suggest yet another distinction, namely the one between novel and established replacements. In the former, humanely killed new animals are used to provide cells, tissues, or organs. In established replacements, permanent cultures of animal cells (cell lines) or tissues are used. An advantage with established replacements is that the total number of animals used does not increase. This means that it is in line with the principle of reduction.

The potential for replacement of animals varies among different fields. Most progress has been made in toxicity testing. In biomedical research replacement appears more difficult (Nuffield Council on Bioethics, 2005, pp. 194-195). The central problem is that molecular, cell, tissue, or organ models are highly simplified compared with whole animals (Smith, 2001).

The report of the Nuffield Council on Bioethics states that it is increasingly obvious that replacement should be discussed on a case-by-case basis instead of in general terms, otherwise progress can hardly be expected (Nuffield Council on Bioethics, 2005, p. 190). I agree. The options for replacement depend on the specific scientific question to be answered. Different questions may suggest different methods. An "alternative" method may not be possible to use in answering the same question as the animal experiment is designed to answer. This case-by-case approach is in line with the casuistic approach of the weak human priority position proposed earlier.

There exist many barriers to replacements. In the report from the Nuffield Council on Bioethics several of these barriers are presented and some proposals are given for how these can be overcome.

Scientific obstacles are factors such as the diversity of cell types and tissues, the complex interaction of cells and tissues, and the influence of tissue organization on cellular environment. In order to overcome some of these obstacles more scientific research into alternative methods is necessary (Nuffield Council on Bioethics, 2005, p. 196).

Non-scientific barriers include, for example, the reluctance of regulatory authorities to depart from traditional methods, insufficient funding for research on alternative methods, lack of availability of information about replacements, insufficient integration of *in vitro* and *in vivo* research, and conservative attitudes among scientists. Measures need to be taken with regard to all these aspects (Nuffield Council on Bioethics, 2005, pp. 196-199).

B. Reduction

Replacement and reduction both concern the number of animals used in experiments. Why is the number of animals ethically relevant? Singer's utilitarian answer would be that pain is ethically relevant and that pain can be aggregated. Many animals feeling pain count higher than does one animal feeling pain (Singer, 1995). But can pain truly be aggregated? Richard Ryder, among others, have questioned this; individual pain is the only thing that counts. Only individual animals can feel pain—an aggregate cannot. The morally significant measure of pain in a group of animals is the maximum pain felt by any one of them. This implies that Ryder criticizes the principle of reduction. The goal should be not so much to reduce the total number of animals used, but to reduce the pain felt by each individual animal. Our first moral concern should be to help those who suffer most. When their pain is reduced, new maximum sufferers appear, and so on. In this way, the number of suffering animals may be reduced, but the severity of pain is still more important to reduce than the number of animals (Ryder, 1999). Ryder has a point, although

I cannot accept his view that the number of animals feeling pain is not an ethically relevant aspect of its own. The severity of pain is important, but the total number of animals feeling pain is also crucial. Ryder appears to mean that it would be better if a large number of animals suffers only mildly than if a small number suffers severely. This might hold true in some cases, but not in all. It would depend on, for example, the more precise number of animals involved and the more precise severity of pain. Balancing severity and number is very difficult, and it cannot be carried out in the manner of pure utilitarian aggregation (see Chapter Five).

As we saw in our analysis of the strong human priority prototype, the principles of replacement and reduction were questioned by Cohen, but from a completely different point of view (Cohen, 1994). In his opinion, more—not fewer—animal experiments are needed in order to fight human disease. Only in this way can scientists fulfill their moral obligations to the other members of the human moral community.

A crucial distinction in this context—which is not mentioned by Russell and Burch—is the one between reduction in absolute terms and reduction in relative terms. What is crucial is that the number of vital scientific questions that are answered by each experiment—or, more precisely, per animal—increases. Relative reduction is more important than absolute reduction.

How can the number of animals be reduced in animal experiments? Russell and Burch suggest two main strategies. The first is to reduce variation. Scientists should use more homogeneous stocks. This will allow the use of fewer animals. The second strategy is better statistical analysis (Russell and Burch, 1992, pp. 105–114). In this regard, the Nuffield report mentions a survey of 78 experiments. Of these, over 60% had obvious statistical errors. This suggests the value of consulting statistical experts before carrying out animal experiments. It also suggests the importance of better training of young scientists (Nuffield Council on Bioethics, 2005, p. 207).

Other things can also be done. Literature search is crucial in order to avoid duplication and to learn about better strategies. In addition, a series of pilot experiments may suggest that some full-scale experiments should not be carried out or indicate better ways of conducting them, thus reducing the number of animals used. Harmonization of guidelines for testing in different countries is also vital. In this way the number of animals used in safety and efficacy testing may be reduced (Nuffield Council on Bioethics, 2005, pp. 205, 208–209).

It is vital to link reduction to refinement, as can be seen in Michael Festing's revised definition of reduction:

> The use of fewer animals in each experiment without compromising scientific output and the quality of biomedical research and testing, and without compromising animal welfare (Festing, 1994).

It might be possible to carry out more procedures on each animal and in this way reduce the number of animals used, but this might compromise animal welfare.

Let me finally give an example of successful reduction. It concerns the LD50 test, that is, the lethal dose 50% test. This test was used for a long time to evaluate the oral toxicity of a single dose of a chemical substance. The test was much criticized (see, for example, Singer, 1995, pp. 53–56; Regan, 1983, pp. 370–371). A few of these tests are still in use but several alternative methods have been developed and are used (Nuffield Council on Bioethics, 2005, p. 209).

C. Refinement

Russell and Burch viewed the principle of refinement as a requirement to refine the techniques to minimize animal pain and distress (Russell and Burch, 1992, p. 134). This means that their focus was on the animal experiment as such and on animal pain. Today, it is common to broaden the scope of the principle in both respects. First, refinement includes not only the experiment in a narrow sense but also all other aspects of the life of the laboratory animal such as housing, husbandry, and care. Second, refinement includes not only reduction of suffering but also improvement of welfare (Smith, 2001). In the next chapter I will investigate the concept of animal welfare in more detail. Let us here look at a few practical issues.

According to the report by Nuffield Council on Bioethics, refinement is likely to be the easiest to achieve in the short term, easier than the other two Rs (Nuffield Council on Bioethics, 2005, p. 210). Refinement can be achieved regarding four different aspects: (1) housing, husbandry, and care, (2) experimental procedures, (3) management of pain, and (4) endpoints.

Since laboratory animals spend most of their time in cages and pens, the refinement of housing, husbandry, and care is crucial. From the scientific literature, it is fairly clear what the species-specific needs of the animals are, physiologically and behaviorally. Of central importance is an "enriched" environment, specified for each species and strain. For chickens this means a nest box, for rats social companions, and for pigs rooting materials, just to mention a few examples (Nuffield Council on Bioethics, 2005, pp. 210–211).

Experimental procedures range from blood sampling to major surgery. Refinement can be sought in all varieties. The Nuffield report mentions the administration of substances as a type of procedure with particularly great potential for refinement. For instance, it is pointed out that the needles must be as small as possible and that the animals are to be kept calm and held very still (Nuffield Council on Bioethics, 2005, p. 212).

Management of pain is also an aspect in need of refinement. A particular problem in this respect is the fact that many laboratory animals conceal signs of pain, making it difficult to detect signs of mild discomfort or distress. Special training of the personnel might be necessary. Studies in animal welfare can make a difference. An example mentioned in the Nuffield report concerns rats undergoing abdominal surgery. These rats exhibit flank twitching, which indicates that they need more pain relief (Nuffield Council on Bioethics, 2005, pp. 212–213).

At the end of experiments, the animals are commonly euthanized. The scientific reasons are that tissues may need to be used for further analysis or the simple fact that the animals can no longer be used. However, the need for "humane" endpoints is increasingly clear. If animal suffering increases during the experiment, it might reach a point where it is judged to be too severe. A refinement of such endpoints is vital so that early clinical signs can be used as indicators that the animal should be euthanized (Nuffield Council on Bioethics, 2005, pp. 213–214).

D. Taking Responsibility

Finally, I would like to stress that perhaps the most important thing regarding all 3Rs is that scientists take responsibility actively and personally (*cf.* Nordgren, 2001). Regulations are important, but only if scientists take personal responsibility can real progress be made. Individual scientists and research teams need to engage in the work of implementing the 3Rs.

Five

ANIMAL WELFARE AND ETHICAL BALANCING

The weak human priority prototype implies that human research interests in most cases—but not all—have a higher ethical priority than animal interests. According to this prototype, considerations of animal welfare may have an impact both on *whether* an animal experiment should be carried out and on exactly *how* it should be carried out. This means that considerations of animal welfare are pivotal. However, many different conceptions of animal welfare have been suggested. To further develop the proposal, it is therefore necessary to analyze some of these conceptions.

1. Three Animal Welfare Concerns

In several articles, David Fraser—alone and together with collaborators—has distinguished three different animal welfare concerns (Fraser, 1993; 1995; Duncan and Fraser, 1997; Fraser *et al.*, 1997; Fraser, 2003). One concern is normal biological functioning. Another is absence of negative affective states—for example, pain or suffering—or presence of positive affective states—for example, pleasure or preference-satisfaction. A third concern is natural living. The conceptions of animal welfare proposed in the interdisciplinary discussion express one of these concerns or combine two or three of them. As I use the terms, "concern" is a matter of caring for the animals' welfare (in one sense or another), while "conception" is a matter of the cognitive content. A conception of animal welfare is an interpretation of the concept of animal welfare. People are talking about the same idea or concept but are offering different accounts of this concept hence they have different conceptions.

Which concern we emphasize may have practical implications for the treatment of animals. Sometimes differences in treatment depend on which concern we stress; commonly, concerns overlap. Let me briefly give a few examples of these animal welfare concerns and then exemplify their practical implications, that is, their "cash value" (a recent and much more elaborate philosophical analysis of different animal welfare conceptions can be found in Nordenfelt, 2006).

A. Biological Functioning

A key example of an animal welfare conception focusing on biological functioning is the one proposed by Donald M. Broom. He states:

> The welfare of an individual is its state as regards its attempts to cope with its environment (Broom, 1991, p. 4168).

In Broom's view, welfare is a matter of coping, and "coping means having control of mental and bodily stability" (Broom, 1993, p. 16).

Although Broom's focus is on function, he also relates feeling to welfare:

> The subjective feelings of an animal are an extremely important part of its welfare. Suffering should be recognized and prevented wherever possible (Broom, 1996, p. 26).

Broom also recognizes the importance of normal behavior for welfare:

> An abnormal behaviour might help an individual to cope, but it is still an indicator that the animal's welfare is poorer than that of another animal that does not have as much difficulty in coping (Broom, 1991, p. 4171).

A quite different example of the biological functioning approach is found in C. J. Barnard and J. L. Hurst (1996). They criticize the focus on coping and suggest an evolutionary account in terms of fitness:

> We contend that welfare can be interpreted only in terms of what natural selection has designed an organism to do and how the circumstances impinge on its functional design. Organisms are designed for self-expenditure and the relative importance of self-preservation and survival, and the concomitant investment of time and resources in different activities, varies with life history strategy (Barnard and Hurst, 1996, p. 405).

From an evolutionary point of view, what counts is fitness in terms of gene replication, not coping in terms of preservation of the individual. Individuals are expendable in the process of reproduction.

Barnard and Hurst criticize the subjective experience approach. First, they stress that we cannot generalize our subjective states to other species because they may have completely different subjective experiences or no capacity for subjective experience whatsoever. Second, they point out that suffering-like states should be defined as those outside the range in which the organism is designed to function, not by generalizing from human experience of suffering (Barnard and Hurst, 1996).

B. Feeling

Other scientists have stressed that subjective feelings and suffering are central to concern about animal welfare. Ian Duncan is straightforward, emphasizing that

> neither health nor lack of stress nor fitness is necessary and/or sufficient to conclude that an animal has a good welfare. Welfare is dependent on what animals feel (Duncan, 1993, p. 12).

Marian Stamp Dawkins argues:

> To be concerned about animal welfare is to be concerned with the subjective feelings of animals, particularly the unpleasant subjective feelings of suffering and pain (Dawkins, 1988, p. 209).

It appears that Dawkins stresses the absence of negative feelings (see also Dawkins, 1980). A philosopher who also focuses on feelings in a broad sense is Peter Sandøe. Stressing positive affective states, he defines welfare as "experienced preference-satisfaction" and continues:

> A subject's welfare at a given point in time (t1) is relative to the degree of agreement between what he/it at t1 prefers (is motivated to do, wants, aspires after, hopes for, does not try to avoid, or is not indifferent to getting) and how he/it at t1 sees his/its situation (past, present and future)—the better agreement the greater welfare (Sandøe, 1996, p. 12).

Thus, we see that feeling-based accounts may focus on the absence of negative affective states or the presence of positive affective states. A focus on positive affective states is a more demanding conception, at least if we talk about the ethical implications (see below).

C. Natural Living

Marthe Kiley-Worthington has indicated a concern for natural living:

> If we believe in evolution ... then in order to avoid suffering, it is necessary over a period of time for the animal to perform all the behaviors in its repertoire, because it is all functional; otherwise it would not be there (Kiley-Worthington, 1989, p. 333).

But Kiley-Worthington views the necessity of performing the full behavioral repertoire as only instrumental to avoiding suffering. Feeling instead of natural living appears to be the key aspect of animal welfare.

In his approach to animal welfare, Bernard Rollin also includes feeling (pain and suffering) as a central aspect, but suggests a broader approach, focusing on the concept of animal nature or *telos*:

> It is likely that the emerging social ethic for animals, which borrows key concepts from our consensus ethic for humans and applies them, mutatis mutandis, to animals, will demand from scientists data relevant to a much increased concept of welfare. Not only will welfare mean control of pain and suffering, it will also entail nurturing and fulfillment of the animals' natures, what I call *telos* (Rollin, 1993, p. 48).

Rollin calls this broader and more comprehensive approach an "increased concept of welfare." The choice of the term *telos* indicates that Rollin is inspired by Aristotle.

The complexity of this approach can be seen in the following quotation:

> It is plausible to suggest that happiness resides in the satisfaction of the unique set of needs and interests, physical and psychological, which make up what I have called the *telos*, or nature, of the animal in question. Each animal has a nature which is genetically and environmentally constrained, from which flow certain interests and needs, whose fulfillment or lack of it *matter* to the animal (Rollin, 1989, p. 203).

We see here (1) that animal nature or *telos* is a unique set of needs and interests, (2) that these needs and interests are physical and psychological, (3) that this nature or *telos* is individual instead of species-specific, and (4) that this nature or *telos* is constrained by genes and environment. Moreover, happiness is central and related to the satisfaction of the needs and interests making up animal nature or *telos*. This means that Rollin stresses both subjective ("happiness") and objective ("nature") aspects of welfare.

In his book *The Frankenstein Syndrome: Ethical and Social Issues in the Genetic Engineering of Animals*, Rollin discusses the relation between *telos* and genetics. He states that *telos* is genetically based and environmentally expressed (Rollin, 1995, p. 172). This means that according to Rollin genetic make-up is of particular importance for *telos*. He also indicates that *telos* includes even more than the full repertoire of functional behaviors by stating that, for example, the *telos* of a pig is its pigness (Rollin, 1995, p. 159).

Rollin also maintains that it may be acceptable to change the *telos* by genetic modification in order to make the animal fit the environment (Rollin, 1995, pp. 171–176). This is vital to note, because reference to animal *telos* is often made in arguments against genetic modification.

Another approach—which is explicitly neo-Aristotelian—can be found in a recent book by the philosopher Martha Nussbaum (Nussbaum, 2006). Nussbaum has for a long time argued for a "capabilities approach" in the sphere of human justice. She now applies this approach to non-human animals. It should be noted that the capabilities approach is a theory of justice and not of animal welfare. However, the conception of animal welfare that appears to be presupposed in her approach comes close to the natural living approach. The focus is on species-specific flourishing, a typically Aristotelian concept. Nussbaum states:

> In short, the species norm (duly evaluated) tells us what the appropriate benchmark is for judging whether a given creature has decent opportunities for flourishing. The same thing goes for nonhuman animals: in each case, what is wanted is a species-specific account of central capabilities (which may include particular interspecies relationships, such as the traditional relationship between the dog and the human), and then a commitment to bring members of that species up to that norm, even if special obstacles lies in the way of that (Nussbaum, 2006, p. 365).

The "species norm" is based on judgment. Nussbaum explicitly clarifies that

> the species norm is evaluative ... it does not simply read off norms from the way nature actually is. But once we have judged that a capability is essential for a life with human dignity, we have a very strong moral reason for promoting its flourishing and removing obstacles to it (Nussbaum, 2006, p. 347).

Although Nussbaum here talks about human beings, the same holds true for non-human animals.

Nussbaum stresses that ethically relevant differences exist between species regarding capabilities and flourishing:

> Because the capabilities approach finds ethical significance in the unfolding and flourishing of basic (innate) capabilities—those that are evaluated as both good and central—it will also find harm in the thwarting or blighting of those capabilities. More complex forms of life have more and more complex (good) capabilities to be blighted, so they can suffer more and different types of harm (Nussbaum, 2006, p. 361).

Central to Nussbaum's approach is that animal welfare goes beyond mere pain and suffering and includes all sorts of capabilities (Nussbaum, 2006, pp. 357–366, 384–388). She applies her "capabilities list" from the human sphere to animals: (1) life, (2) bodily health, (3) bodily integrity, (4) senses, imagination and thought, (5) emotions, (6) practical reason, (7) affiliation, (8) interspecies relationships, (9) play, and (10) control over the environment. Although she is aware of and discusses the problems of applying this list to animals, she believes that it will offer good guidance for policies regarding animals (Nussbaum, 2006, pp. 392–401).

In sum, we find at least four different versions of the "natural living" conception. The first focuses on natural (or normal) behavior. For example, a hen works for a perch and a sand-bath and a sow in hot weather wants to wallow. Such behaviors can be more or less exhibited also in quite constrained circumstances, for example, in a farmhouse. The second view is more encompassing and focuses on natural living in a fuller sense. It concerns—more or less—the animal's life as a whole and consists in leading this life in the species' natural (wild) environment (habitat). The third view has a somewhat different focus, namely animal nature. This nature is genetically based and includes more than the full repertoire of natural behaviors, for example, "pigness." The fourth view highlights species-specific innate capabilities and flourishing.

D. Commonly Overlapping ...

Commonly considerable overlap exists between the three animal welfare concerns. The welfare of an animal may be assessed to be good or bad whatever animal welfare conception we apply. For example, if an animal has a life-threatening disease such as cancer, or has a severely injured leg, this would imply bad welfare on all accounts, although the reasons might differ. According to conceptions focusing on absence of suffering, the pain would imply poor welfare in both cases. Conceptions focusing on functioning, on the other hand, would imply that organs with cancer are dysfunctional and that a severely injured leg prevents normal mobility. Finally, conceptions focusing on natural living would imply that such living in both cases is hindered.

E. ... But Sometimes in Conflict

Sometimes the practical implications of the three concerns differ, leading to conflicting assessments of welfare. This is often acknowledged in the debate. For example, an animal may have poor welfare in terms of biological functioning but acceptable welfare in terms of feeling. It may be injured, have an impaired immune system, have increased susceptibility to disease, be unable to reproduce, or be at risk of premature death; despite this it may not feel pain (Broom, 1991).

There may also be practical differences between conceptions focusing on functioning and those focusing on natural living. For example, biological functioning may be promoted by protecting farm animals from the hardships of nature. The animals are provided with shelter, nutrition, and protection from predators. A focus on natural living, on the other hand, may imply more freedom in terms of avoidance of the crowded housing and rough handling of conventional farming (Fraser, 1995).

Conceptions based on natural living may be in conflict with those based on feeling. Take, for example, a veal calf confined in a crate and fed a low-iron liquid diet (Tannenbaum, 1991). From the natural living perspective, its welfare may be assessed to be low, since it has no ability to move and socialize. From the subjective experience perspective, the welfare may be acceptable since the animal is not experiencing pain. Another example is given by Rollin. He argues that it would be wrong to capture a gazelle, tiger, or eagle, and keep them in cages. The reason is that this would violate the "nature" of the animal. This would be wrong, according to Rollin, even if the animal did not feel any pain and even if it experienced much pleasure (Rollin, 1981, pp. 34–35).

Let me finally point out a possible difference between conceptions focusing on absence of negative feelings and those focusing on presence of positive ones. An example is laboratory animals. Given a conception of the negative type, the goal may merely be alleviation or prevention of pain. Given a conception of the positive type, the goal may also be to allow animals to enjoy pleasures that are not necessary for alleviation or prevention of pain. This can be promoted, for example, by providing enriched environments (Tannenbaum, 2001).

2. A Comprehensive Approach

Fraser criticizes restrictive views that focus on only one or two of the three animal welfare concerns (note that below I will for the sake of simplicity refer to Fraser only and ignore the co-authors of some of his papers):

> We suggest, instead, that if animal welfare research is to address major ethical concerns about the quality of life of animals, then the conception of animal welfare used by scientists needs to reflect the full range of major ethical concerns extant in society (Fraser *et al.*, 1997, p. 202).

His alternative is "an integrative model" (Fraser *et al.*, 1997, p. 199). He argues:

> Scientific research on "animal welfare" began because of ethical concerns over the quality of life of animals, and the public looks to animal welfare research for guidance regarding these concerns. The conception of animal welfare used by scientists must relate closely to these ethical concerns if the orientation of the research and the interpretation of the findings is to address them successfully. At least three overlapping ethical concerns are commonly expressed regarding the quality of life of animals (Fraser *et al.*, 1997, p. 187).

As far as I can understand, two different theses can be found in these quotations:

(1) a conceptual thesis: an adequate conception of animal welfare should reflect three ethical concerns: functioning, feeling, and natural living,

(2) an ethical thesis: an adequate animal ethic should include three inherently important animal welfare concerns: functioning, feeling, and natural living.

The first thesis is explicitly stated. It means that all three concerns should be considered inherently important, not one or two of them merely instrumentally important (Fraser, 2003). The second thesis is only implicit. All three concerns are ethically legitimate goals of human action. The conceptual thesis (1) presupposes the ethical thesis (2). Only if the ethical thesis is legitimate is the conceptual thesis legitimate.

Fraser's argument for this comprehensive approach to animal welfare is that it reflects better than other approaches the major ethical concerns regarding animal welfare that are "extant in society" (Fraser *et al.*, 1997, p. 202). These concerns are sincere moral intuitions among the general public that express a caring attitude toward animals. Animal welfare science has to take all three ethical concerns seriously in order to respond to the general public and retain its confidence. Fraser claims that "the public looks to animal welfare research for guidance regarding these concerns" (Fraser *et al.*, 1997, p. 187).

If we are to understand Fraser's argument, it must be put into context. He is writing in scientific journals in which the concept of animal welfare and its possible value-laden character has been heavily debated for years. Sometimes this discussion has been quite technical. Fraser wants to broaden the perspective and asks which concerns basically motivate animal welfare science. He maintains that these concerns are ethical and that they are expressed not only by scientists but also by the general public. To support this statement, he quotes several scientists and lay persons (Fraser *et al.*, 1997).

I am prepared to accept Fraser's comprehensive approach, provided that certain conceptual and ethical problems are handled in a satisfying way. Let us take a look at these problems.

3. Conceptual Implications

A. Conflicting and Incommensurable Components

The comprehensive approach states that an adequate conception of animal welfare consists of three components that reflect three ethical concerns—functioning, feeling, and natural living—and that are seen as inherently important. This gives rise to two problems. The first problem is that the three components may sometimes be in conflict. I gave a few examples above. The second is that the three components are incommensurable, that is, they cannot be measured by a common unit of measurement. These problems imply that it is impossible to give a definition of the comprehensive concept of animal welfare by providing necessary and sufficient conditions. No true conceptual integration is possible. Therefore, Fraser's designation—"an integrative model"—does not seem quite appropriate. The term "comprehensive" is better, since it merely suggests a conglomerate of different components.

These conceptual problems raise an even more basic issue: can a concept with these properties be theoretically acceptable? The cognitive linguistics of George Lakoff and Mark Johnson—mentioned earlier—can shed light on this issue. Lakoff and Johnson have carried out empirical investigations that indicate that many—and, in their opinion, most—concepts cannot be defined by providing necessary and sufficient conditions. The reason is that the concepts are metaphors and have prototype structure (Lakoff and Johnson, 1980; Johnson, 1993; Lakoff and Johnson, 1999; both notions have been explained above).

The concept of welfare is fundamentally a metaphor. Lexically, the word "welfare" means "fare well." This suggests that welfare is a metaphor of traveling. This traveling has positive value. To fare well is to travel "safely and conveniently." These terms are also metaphors. To travel "safely" means to travel with no accidents or no serious risk for accidents. To travel "conveniently" means to travel smoothly and without hindrances.

The concept of welfare has prototype structure. First, each conceptual component—functioning, feeling, and natural living—has prototype structure. Clear, commonly accepted examples exist of functioning well, feeling well, and living naturally, respectively. We also find non-prototypical instances that are unclear and disputed. For example, the borderline between functioning well and not functioning well is vague, and scientists may draw the line differently. Second, the comprehensive concept also has prototype structure. Prototypes are, for example, the cases of overlap given above. Non-prototypical cases are the examples of conflict between the conceptual components.

The fact that the comprehensive conception of animal welfare cannot be defined by necessary and sufficient conditions but is a metaphorical concept based on three incommensurable prototypes may be criticized as theoretically unsatisfying. However, the metaphorical character and prototypical structure are shared by many other—and perhaps most—concepts. Moreover, the comprehensive conception has the advantage of reflecting all three ethical concerns. These concerns are expressed not only in the academic discussion but in society at large.

B. A Value-Laden Concept

The concept of animal welfare is value-laden, which is indicated by its basic metaphor of "faring well," mentioned above. Fraser also stresses this, talking about "ethical concerns" (Fraser *et al.*, 1997, p. 189). However, he goes beyond mere description and makes a normative proposal, stressing that an acceptable scientific conception of animal welfare "needs to reflect the full range of major ethical concerns" (Fraser *et al.*, 1997, p. 202) and "must relate closely to these ethical concerns" (Fraser *et al.*, 1997, p. 187).

Many other participants in the debate share Fraser's recognition of the value-laden nature of the animal welfare concept. It is stressed, for example, by Rollin:

> A main component of 20th century scientific ideology is the view that science is "value-free." This notion has dominated the view of animal welfare in the emerging field of animal welfare science. Science, however, is neither value-free in general, nor ethics-free in particular. The value-laden nature of the concept of "animal welfare" is clear, and even what information is considered to count as facts is structured by valuational presuppositions (Rollin, 1993, p. 44).

It is not enough to point out the value-laden character of the animal welfare concept. The nature and role of values must be clarified more precisely. The values involved can be of many different types: ethical, preferential, epistemological, methodological, and so on. Fraser appears to focus primarily on ethical values, as do F. R. Stafleu *et al.*, who pinpoint "the moral aspect of welfare" (Stafleu *et al.*, 1996, p. 233), and Tannenbaum, who states that "the very concept of animal welfare—what ordinary people and scientists mean by the term welfare—includes an ethical component" (Tannenbaum, 1991, p. 1366).

Ethical values may enter animal welfare science in several different ways. Fraser has concentrated on the three ethical concerns about animal welfare that initially motivates animal welfare science. Because of these concerns, scientists make choices regarding "definitions, parameters, levels to be studied, etc." (Stafleu *et al.*, 1996, p. 232). These choices involve different types of values, including ethical ones. When the welfare assessment of a particular animal has been carried out, the ethical problem arises of how to treat the animal on the basis of these results. An even more general ethical problem arises: Are there uses of animals in human society that imply too badly deficient animal welfare to be ethically acceptable? In this book, the focus is on animal experimentation. A key problem is whether some experiments should not be carried out due to animal welfare concerns.

Given this "ethics-laden" nature of the concept of animal welfare, let us take a closer look at the ethical problems raised by the comprehensive approach to animal welfare.

4. Ethical Implications

A. Balancing Conflicting Animal Welfare Concerns

According to the ethical thesis of the comprehensive approach, all three ethical concerns—functioning, feeling, natural living—are legitimate ethical concerns. This raises the problem of how to balance these ethical concerns when they conflict. What is the order of priority of the three inherent animal welfare concerns: functioning, feeling, and natural living? Fraser explicitly recognizes this problem. He talks about "weighing conflicting but incommensurable variables" (Fraser, 2003, p. 441), but he does not clarify how to solve the problem. His aim is only to show that science cannot solve it and that ethical deliberation is needed.

Two different ways of solving the problem exist. One is to make a lexical ordering, for example by stating that feeling is always most important and functioning and natural living always have lower priority. Another option is contextual ordering, or to state that the respective priority of the three ethical animal welfare concerns may vary from one context to another. Sometimes functioning is most important, sometimes feeling, and sometimes natural living. We prioritize differently because of the special relations of the animals to human beings, that is, because of different uses or functions of animals in human society.

Fraser appears to favor contextual ordering. His use of the term "weighing" (Fraser, 2003, p. 441) in characterizing the ethical problem is a weak indication of this. I also support contextual ordering. Lexical ordering is not sufficiently sensitive to the variety of contexts, such as those in which animals are used in human society and those in which they are not used, that is, animals in the wilderness that are not "used" for hunting.

When we choose to accept a kind of use of animals in society, we by implication also set certain priorities among animal welfare concerns. If we, for example, accept—on certain conditions—the use of animals in farming, this choice implies that the ethical concern for natural living receives a low priority, since the animals are to be kept in confined areas (more or less spacious). Conversely, the ethical concerns for functioning well or feeling well get higher priority. We want our farm animals to function well and feel well. Similarly, if we accept—on certain conditions—the use of animals in scientific experimentation, we also by implication give the ethical concern for natural living a low priority, because the animals have to be kept in cages for scientific reasons, namely, control. And conversely the ethical concerns for functioning well and feeling well receive higher priority.

Consider wild animals as compared with farm animals and laboratory animals. In the context of wilderness, natural living is the most important animal welfare aspect, although we may sometimes euthanize injured wild animals to put an end to their pain, and even help animal offspring whose parents have died to develop into well-functioning adults. However, we do not on a regular basis make strong efforts to look up and actively help wild animals to feel well or function well.

Even choices regarding more particular uses of animals may also have implications for which ethical concerns are considered the most important. If we, for example, decide to use animals in research as disease models or in knock-out studies of gene function, the ethical concern for functioning well in terms of health would, by implication, have low priority, because it is precisely effects in terms of ill health that are sought. The ethical concern for feeling well would then be given higher priority. Although we seek clinical effects, we want to keep the pain at a minimum.

We may object to the idea that we first determine which uses of animals are ethically acceptable and then by implication or in addition determine which animal welfare concern should be prioritized. Should we not instead ask if the animals' use in society is itself acceptable in animal welfare terms? Here it should be noted that the choices we make concerning the use of animals may be based on different kinds of ethical considerations. These considerations may be animal welfare considerations, but they may also be considerations of another type, for example animal rights considerations or considerations stressing the importance of human interests. If we choose to criticize a particular kind of use of animals, this can be done on the basis of animal welfare considerations (as in Singer) but also by referring to animal rights (as in Regan). And the criticism might be categorical or non-categorical, that is, be against all such uses or only against some or most of them. If we choose to accept a particular kind of use of animals, then that choice may be justified by reference to human benefit in addition to animal welfare considerations (as in Cohen and Midgley) or by human benefit only (as in Carruthers). This acceptance of a particular use may also be categorical or non-categorical.

To the extent that animal welfare is taken into account in criticism or defence of a particular animal use, the focus may be on different animal welfare aspects: feeling (as in Duncan), functioning (as in Broom) or natural living (as in Rollin). Alternatively, all three animal welfare aspects are accepted and their ordering is considered to be contextual (as Fraser suggests).

Let me here make a short comment on the third of the 3Rs presented in the previous chapter, namely the principle of refinement. This principle was originally understood by Russell and Burch as a requirement to refine the techniques to minimize animal pain and distress (Russell and Burch, 1992, p. 134). What is apparent here is the presupposed subjective approach to animal welfare. The focus is on animal feelings of pain and distress, not on functioning or natural living, and the focus is exclusively on avoiding negative affective states, not on promoting positive ones. But positive feelings are also an important concern and are generally considered to be so today (Smith, 2001). Animals can be stimulated to experience such feelings by an enriched environment in terms of wheels, piles of hay, and so on. This is an important concern in housing, husbandry, and care before the experiment (in a narrow sense) is carried out, but in the experimental situation keeping the pain at a minimum is the primary concern.

B. The Comprehensive Approach and Casuistry

Fraser's comprehensive approach has a serious weakness, namely that the ethical concerns are not related to ethical theory. In my opinion, the best option is to combine the approach with the casuistic view suggested in the previous chapter. Let me explain.

The comprehensive approach involves three ethical concerns, namely functioning well, feeling well, and natural living. One of these fits utilitarianism, namely feeling well. The goal is to avoid suffering and promote pleasure (hedonistic utilitarianism) or to achieve experienced preference-satisfaction (preference utilitarianism). In both types of utilitarianism, the subjective mental states of the animals are considered inherently important. According to the comprehensive approach, objective states in terms of biological functioning and natural living are also inherently important. Thus, this approach combines utilitarian and non-utilitarian animal welfare concerns.

This mixture of subjective (feelings-based) and objective (functioning-based and natural living-based) welfare concerns is not without problems. Fraser recognizes that the concerns may be difficult to combine and calls them "incommensurable variables" (Fraser, 2003, p. 441), but I find no indication that he recognizes the relative oddness of this mixture from the perspective of established ethical theory. Utilitarian and non-utilitarian welfare concerns are commonly viewed as distinct (Appleby and Sandøe, 2002; Sumner, 1996).

In animal welfare legislation, however, we sometimes find such combinations. For example, in my own country—Sweden—the Animal Welfare Act (1988 (with later revisions)) states that "animals shall be treated well and shall be protected from unnecessary suffering and disease" (Section 2 (1)) and that "animals shall be accommodated and handled in an environment that is appropriate for animals and in such a way as to promote their health and permit natural behavior" (Section 4 (1)). In this legislative framework, traces of all three animal welfare concerns can be found. The focus on "suffering" shows a concern for animal feeling. The statements about "disease" and "health" suggest a concern for biological functioning, and the statement regarding "natural behavior" indicates a concern for natural living. In animal experimentation, feeling appears to be the primary aspect. With regard to the ethics committee on animal experimentation, the Animal Welfare Ordinance (1988 (with later revisions)) states that "when considering specific cases the committee shall weigh the importance of the experiment against the suffering inflicted on the animal" (Section 49 (1)).

Another example of a combination of different welfare concerns is found in the list of the "five freedoms" issued in the United Kingdom by the Farm Animal Welfare Council:

1. Freedom from thirst, hunger and malnutrition—by providing ready access to fresh water and a diet to maintain full health and vigour.
2. Freedom from discomfort—by providing a suitable environment including shelter and a comfortable resting area.
3. Freedom from pain, injury and disease—by prevention or rapid diagnosis and treatment.
4. Freedom to express normal behaviour—by providing sufficient space, proper facilities and company of the animal's own kind.

5. Freedom from fear and distress—by ensuring conditions which prevent mental suffering (FAWC, 2009).

Two things should be noted here. First, all three animal welfare concerns are alluded to. Much focus is on feeling ("discomfort," "pain," "fear," "distress," and "mental suffering") but also on functioning ("thirst," "hunger," "malnutrition," "health," "vigor," "injury," "disease," "diagnosis," and "treatment") and natural living ("normal behavior," "sufficient space," "company of the animal's own kind," and "suitable environment"). Second, no attempt is made to reduce the concerns to one single most important one. All concerns seem to be important in their own regard.

Another aspect of the comprehensive view is that the incommensurable ethical concerns are to be weighed in cases of conflict. This talk about "weighing" incommensurable variables (Fraser, 2003, p. 441) can be interpreted in terms of a casuistic approach. Weighing is typical of utilitarianism, but not the weighing of "incommensurable variables." Casuistry, on the other hand, is characterized precisely by the contextual weighing of a plurality of divergent ethical concerns (Jonsen and Toulmin, 1988). In particular, the comprehensive ethical approach comes close to Baruch Brody's "pluralistic casuistry," which includes a plurality of legitimate ethical appeals—teleological and deontological—to be balanced on a case-by-case basis (Brody, 1988; Brody, 1998, pp. 197–212). The comprehensive approach also fits my own "imaginative casuistry" very well; it is also pluralistic (see Chapter Three; Nordgren, 1998; Nordgren, 2001, pp. 15–49).

The casuistic approach to weighing implies a contextual ordering of conflicting animal welfare concerns. In two of the examples of contextual ordering given above—farming and scientific experiments—it is presupposed that it is ethically acceptable to use animals for the sake of human welfare in these contexts. It is presupposed that it is ethically acceptable for human beings to be involved in these special relations with animals.

This raises—in turn—the question of whether poor welfare could sometimes count as a reason for not using animals in a particular context. According to the weak human priority prototype, this may sometimes be the case. With Midgley, I accept that human beings are involved in special relations with animals, although animal welfare concerns may in some contexts outweigh human welfare concerns, implying that animals should not be used in these contexts (Midgley, 1983, pp. 19–32, 98–113). The acceptance of relational properties as ethically relevant makes it possible to balance different animal welfare concerns differently in different contexts. For example, it is possible to balance them differently regarding wild animals, farm animals, and laboratory animals, as suggested above.

C. Balancing Conflicting Animal Welfare and Human Welfare Concerns

Within the comprehensive approach not only problems of balancing conflicting animal welfare concerns may arise, but also problems of balancing conflicting animal welfare concerns and human welfare concerns. In cases of conflict, should human welfare—regardless of how it is conceived—be assigned higher priority than animal welfare? Which uses of animals—if any—are ethically acceptable? These issues are by no means unique to the comprehensive approach. They are basic problems for all types of animal ethics. This problem is the same even if only one or two of the three animal welfare concerns are embraced.

The five ethical prototypes of animal experimentation discussed in previous chapters suggest different ways of handling this problem. As in the case of balancing conflicting animal welfare concerns, we find basically two ways of balancing conflicting animal welfare and human welfare concerns, namely lexical and contextual ordering.

The strong human priority prototype of Cohen proposes a lexical ordering (Cohen 1994; Cohen in Cohen and Regan 2001). All uses of animals contributing to human welfare are acceptable, although animal welfare should always be taken seriously.

At first glance, the human dominion prototype and the animal rights prototype may also be said to suggest lexical ordering, but on closer inspection the problem of balancing does not arise at all or only very rarely.

According to the human dominion prototype suggested by Carruthers (1992), we have no direct duties regarding animal welfare, and therefore no problem of balancing conflicting animal welfare and human welfare concerns arises.

Neither will there commonly arise any problem of balancing in Regan's animal rights prototype (Regan, 1983; Regan in Cohen and Regan, 2001). The reason is that human welfare concerns should in principle never be allowed to interfere with animal welfare concerns. Only in "lifeboat cases" where the lives of both human beings and animals are threatened, may there occur a problem of balancing. In such cases, human welfare may outweigh animal welfare, but such cases are very rare. However, in these very rare cases it appears to be a matter of contextual ordering of welfare concerns instead of lexical.

The prototype of equal consideration of interests defended by Singer (1993a; 1995) suggests contextual ordering, since this utilitarian approach focuses on individual acts and the facts of the particular situation. The outcome of ordering is likely to be that human welfare concerns only rarely outweigh animal welfare concerns in cases of conflict. Most human uses of animals involving low animal welfare in terms of feelings of suffering are unacceptable.

The weak human priority prototype of Midgley (1983) also implies contextual ordering. The outcome of prioritizing conflicting animal welfare and human welfare concerns is likely to be that human welfare concerns commonly—but not always—outweigh animal welfare concerns. Only uses of animals contributing to vital human welfare concerns are acceptable. Taking animal welfare concerns seriously may prohibit some uses of animals in some contexts.

In sum, we have three different views of how to prioritize conflicting animal welfare and human welfare concerns: "strong human priority" proposing lexical ordering, and "weak human priority" and "equal consideration" proposing contextual ordering. I have argued in favor of the weak human priority prototype and consequently I support contextual ordering according to this approach. Human welfare concerns often—but not always—outweigh animal welfare concerns. The problem is which human welfare concerns are sufficiently vital. As argued in the previous chapter, uses of animals for research purposes are often to be considered vital.

Let me finally restate that Midgley does not propose a comprehensive view of animal welfare. She appears to understand welfare primarily in terms of feelings (Midgley, 1983, p. 96). In this book I propose that Midgley's approach to animal ethics is to be supplemented with the comprehensive animal welfare approach of Fraser.

It is enlightening to turn the issue the other way around and ask how Fraser evaluates Midgley. In a paper where Fraser comments on various types of animal ethics, we get a clear indication of this. Fraser criticizes a category of philosophers, which he calls "type 1 philosophers." These philosophers do not take animal welfare science seriously enough. His key examples are Tom Regan and Peter Singer. Fraser is critical toward six aspects of this kind of animal ethics:

> It tended (1) to focus only on the level of the individual rather than making some decisions at the level of the population, ecosystem or species, (2) to advocate single ethical principles rather than balancing conflicting principles, (3) to ignore or dismiss traditional ethics based on care, responsibility, and community with animals, (4) to seek solutions through ethical theory with little recourse to empirical knowledge, (5) to lump diverse taxonomic groups into single moral categories, and (6) to propose wholesale solutions to diverse animal use practices (Fraser, 1999, p. 171).

By contrast, Fraser is positive toward "type 2 philosophers." These philosophers propose approaches to animal ethics that are "compatible" with animal welfare science. Some of this work "attaches value to traditional care for and community with animals" (Fraser, 1999, p. 171). Fraser stresses that this type of animal ethics should be further investigated, stating that "we need better developed theories that articulate the ethical significance of care and community involving other species" (Fraser, 1999, p. 186). One of the philosophers that Fraser mentions as a proponent of this kind of a view is Midgley. Therefore, it appears that out of the five prototypes presented in Chapter Two, Fraser comes closest to Midgley's, although he does not develop a theory of his own.

Let me highlight two particular aspects of Fraser's incompletely developed view. The first concerns contextual ordering in relation to different animal species. As mentioned above, Fraser criticizes the tendency of some animal ethicists to "lump diverse taxonomic groups into single moral categories" (Fraser, 1999, p. 171). He exemplifies this with Regan who talks about "subjects-of-a-life," referring to at least all mentally normal mammals one year old or more (Regan, 1983). Fraser admits that in theory Singer is more discriminating, making a distinction between sentient and non-sentient beings and between persons and non-persons. In practice, however, Singer focuses mainly on the aggregate and recommends, for example, consumers not to buy products that have been tested on "animals" (Singer, 1995, p. 94; Fraser, 1999, p. 175). Fraser, on the other hand, stresses the importance of taking into account empirical knowledge about different species obtained by animal welfare science (Fraser, 1999, p. 176). For example, he mentions with approval that "scientists have proposed the use of domestic pigs rather than primates ..., and fish rather than birds or mammals ..., as more ethically acceptable ways of conducting certain types of animal research" (Fraser, 1999, p. 175).

The second aspect concerns contextual ordering in relation to different animal uses or functions. Fraser criticizes the tendency of some animal ethicists "to lump diverse animal use practices under broad headings such as 'commercial animal agriculture,' and to advocate extremely general remedies for extremely complex situations" (Fraser, 1999, p. 176). Regan is a particularly clear target of this criticism, holding for example that commercial animal production should cease entirely (Regan, 1983, pp. 330–353). Fraser recognizes that in theory Singer is also in this respect more discriminating. But in practice Singer rejects certain broad categories of animal use in totality. One example, pointed out by Fraser, is Singer's view that the meat available in food stores comes from animals that have not been treated with any real consideration at all while being reared (Singer, 1995, p. 160; Fraser, 1999, p. 176). By contrast, Fraser talks about "discriminating between good and bad animal use practices" and stresses "the importance of empirical analysis" in carrying out such a discrimination (Fraser, 1999, p. 171). Certain particular uses of animals are ethically acceptable, while others are not.

In conclusion, Fraser's comprehensive conceptual and ethical approach to animal welfare fits very well with Midgley's type of animal ethics.

5. Animal Sentience

One of the components of this comprehensive conception of animal welfare is feeling. I have argued that it is the most important animal welfare concern in animal experimentation. This component requires further clarification.

We saw in Chapter Two that the proponent of the human dominion prototype, Carruthers, in a speculative chapter of his book stated tentatively that animals are unconscious (Carruthers, 1992, pp. 186, 192–193). In particular, he argued that "if animals are incapable of thinking about their own acts of thinking, then their pains must all be non-conscious ones" (Carruthers, 1992, p. 189). On the other hand, the proponents of the other four prototypes—Cohen, Midgley, Singer, and Regan—all stress that many animals are sentient and can feel pain. I have indicated that Carruthers' speculations have no scientific ground. I will now elaborate on different aspects of animal sentience (without special reference to Carruthers). As a starting point I will take the analysis of David DeGrazia in his book *Taking Animals Seriously: Mental Life and Moral Status* (1996). This book makes many clarifying points concerning animal sentience and consciousness, but I will make a few critical remarks and also criticize DeGrazia's view of animal welfare. In addition, I will give a few constructive proposals.

A. Nociception

The proper starting point for a discussion of animal pain is nociception. Nociception is the activity of nociceptors. Nociceptors detect stimuli that could be tissue-damaging, for example, cold, heat, pressure, and cutting. Nociception is not a mental state. It does not in itself involve pain. But it is the first stage in a process that often includes pain. The nociceptors fire nervous impulses along axons, but they do so only over a particular threshold of intensity. Below that level the nociceptors do not respond to stimuli. DeGrazia gives the example of the paraplegic who touches a hot iron. She does not feel any pain, because of the damage of her spinal cord, but nevertheless she withdraws the foot (DeGrazia, 1996, p. 99). Nociception should be distinguished from other types of responsiveness to stimuli, for example a plant's movement of leaves toward light. According to DeGrazia, nociceptors can be found in all vertebrates and possibly also in cephalopods, although great variation exists in the details. Their presence indicates that an animal may have the ability to feel pain, although it is not a mental state in itself (DeGrazia, 1996, pp. 99–100).

B. Pain

What is pain? DeGrazia proposes the following definition:

> Pain [is] ... an unpleasant sensory experience typically associated with actual or potential tissue damage (DeGrazia, 1996, p. 107).

Several things should be noted here. First, pain is viewed as an experience. It is a mental state, a form of consciousness. DeGrazia does not try to provide a non-circular definition of what consciousness is. Consciousness appears to be "a basic, irreducible concept within our conceptual scheme" (DeGrazia, 1996, p. 101). Second, pain has two components, a sensory component and an affective component. Third, DeGrazia stresses the link between pain and nociception by referring to tissue damage. In this way pain can be distinguished from other unpleasant experiences such as distress and anxiety. For my purposes in this book, DeGrazia's definition appears quite appropriate.

The scientific evidence for pain in animals DeGrazia gathers from three sources. In each case he compares the findings with human beings (DeGrazia, 1996, pp. 105–115).

The first source is ethology. Many animals behave as if they feel pain, but the data must be scrutinized critically. Similar to human beings, many animals exhibit avoidance or escape behavior, for example withdrawing a body part from a hot object. Since this may also be seen in insects, it might be the case that this behavior sometimes is merely reflexive or nociceptive. Vertebrates and cephalopods show ability of adaptation and learning with regard to stimuli associated with pain in human beings. This supports the hypothesis that they do feel pain.

The second source is physiology. Many animals have a neural machinery similar to human beings. The basic neurophysiology of nociception is similar in vertebrates and cephalopods, although the details differ. Anesthesia and analgesia control pain across these species.

Third, from the perspective of evolutionary theory we have reason to believe that animals that are closely related to human beings may have a similar ability to feel pain. In human beings the ability to feel pain is highly functional, and there might be a selective pressure toward the evolution of avoidance and escape behaviors in many related species.

This evidence might seem convincing, but in order to indicate that a dispute is still going on among biologists regarding exactly which species have the ability to feel pain, let me give the example of fish, which has recently been subjected to intensified discussion. In a review article, James Rose argues that the nervous system of fish is very different from that of mammals. Therefore, fish lack the brain structures necessary for generating the consciousness required for experiencing pain (Rose, 2002). On the other hand, K. P. Chandroo *et al.* criticize Rose for not providing sufficient evidence to be able to exclude the possibility that the fish brain could generate some kind of consciousness. Other parts of the brain than the neo-cortex might be important in this respect (Chandroo *et al.*, 2004).

I conclude that we have good reasons to believe that all vertebrates—mammals, birds, fish—and also cephalopods have the ability to feel pain. But we may distinguish prototypical examples and non-prototypical ones. Primates belong to the former category but also the most common species used in animal experimentation, namely mouse and rat. Non-prototypical and more disputed examples are fish and even more so cephalopods.

How do these animals experience pain? This question is not discussed by DeGrazia, but as far as I can see a comment by Barnard and Hurst is relevant in this regard. We saw above that they stress that we cannot generalize our subjective states to other species because they may have completely different subjective experiences or no capacity for subjective experience whatsoever (Barnard and Hurst, 1996). This is an important reminder. It is also vital not to overstate the difficulty of understanding how animals feel. There appears to be a tendency in Barnard and Hurst to go too far into skepticism. Anthropomorphism is a risk, but pain appears to be a better candidate than all other subjective states for being a state that is fairly similar across many species. The reason is that nociception—the neurological basis for feeling pain—is quite similar across species, despite differences in details. This fact indicates the risk of underestimating the similarity between human and animal experience of pain.

How should this limited understanding of how animals experience pain be handled from an ethical point of view? Although understanding more precisely how animals feel pain would be valuable, it does not seem necessary for responsible ethical deliberation. The fact that many animals feel pain—regardless of how they do it—is what is ethically crucial. Moral imagination is not the ability to feel what animals feel. It is the ability to recognize the ethical relevance of animal pain. Moreover, even if we stress the animal ability to feel pain very strongly—at the risk of overstating the similarity of human and animal experience of pain—we must remember that in animal experimentation animal pain has to be balanced against the expected human benefit. As argued above, if the expected human benefit is high, it might nevertheless outweigh severe animal pain.

C. Distress and Suffering

"Sentience" is a common term in the animal ethics debate, which covers the ability to feel pain, but feeling pain is only one aspect of sentience. Other negative affective states are distress and suffering. DeGrazia understands distress in the following way:

> Distress is a typically unpleasant emotional response to the perception of environmental challenges or to equilibrium-disrupting internal stimuli (DeGrazia 1996, p. 117).

This analysis has several important points. Distress is considered to be an emotional response to certain perceptions. These perceptions may be due not only to external stimuli but also to internal ones. With regard to external stimuli, distress—in distinction to pain—is not a reaction to potential tissue damage but to "challenges" in the environment to which the animal is unable to adapt or has difficulty adapting to. The internal stimuli provoking the unpleasant emotional response are characterized by being "equilibrium-disrupting." It is not quite clear what this means, but one interpretation is that they are perceptions of lacking ability to handle challenges.

DeGrazia suggests the following definition of "suffering":

> Suffering is a highly unpleasant emotional state associated with more-than-minimal pain or distress (DeGrazia, 1996, p. 116).

We see here that suffering—according to DeGrazia—is an emotional state. This emotional state is associated with but not identical to more-than-minimal pain or distress. This suggests that it might be possible to experience more-than-minimal pain or distress without suffering. DeGrazia argues that most or all vertebrates and possibly some invertebrates can suffer (DeGrazia, 1996, p. 123).

D. Pleasure and Happiness

Sentience may also include positive affective states, not only negative ones. In his discussion of pleasure, DeGrazia distinguishes two different models, the feeling model (proposed by Bentham) and the attitude model (defended by Sidgwick). According to the feeling model, it is contingent whether or not we like pleasures. According to the attitude model, pleasures are necessarily liked. DeGrazia supports the attitude model. He quotes Sidgwick's definition of pleasure:

> Let, then, pleasure be defined as feeling which the sentient individual at the time of feeling it implicitly or explicitly apprehends to be desirable (Sidgwick quoted in DeGrazia, 1996, p. 124).

A related concept is enjoyment, which DeGrazia understands as "an all-things-considered endorsement of, or preference for, an experience." I find no problem with the term "preference," but "endorsement" appears to require too much ability to make an assessment. According to this definition, enjoyment presupposes an object that we enjoy. This is, on the other hand, not the case with another related concept, namely happiness. Happiness is merely a mood. DeGrazia makes a distinction between feeling happy and being happy. The former is a mental state over a period of time. "Being happy," on the other hand, he understands as

> feeling satisfied or fulfilled by the basic circumstances of one's life, such that one is disposed to endorse or affirm them in terms of one's own priorities (DeGrazia, 1996, p. 126).

Being happy in this sense presupposes the capacity to make an assessment of our life as a whole. This is a quite advanced mental achievement.

According to DeGrazia, most or all vertebrates and possibly some invertebrates can experience pleasure and enjoyment (1996, p. 126). His arguments are quite similar to those he presented with regard to feeling pain: physiological, ethological, and evolutionary-functional. The concept of happiness is more difficult to apply to animals. Many animals can feel happy, but they can hardly be happy, since this would require the ability to make judgments about their lives as wholes (DeGrazia, 1996, p. 127)

DeGrazia's understanding of sentience is very much in line with Singer's. We saw in Chapter Two that Singer used the term sentience as "shorthand for the capacity to suffer or experience enjoyment or happiness"

(Singer, 1993a, p. 58). However, it is radically at odds with Carruthers's account of animal consciousness (DeGrazia, 1996, pp. 112–115).

E. Self-Awareness, Language, and Moral Agency

Sentience must be distinguished from self-consciousness or self-awareness. In contrast to the former, the latter presupposes a concept of a self. We have seen that Singer argues that some sentient animals are "persons" in the sense that they are "self-conscious and rational" (Singer, 1993a, p. 87), and Regan maintains that at least all mammals one year old or more are "subjects-of-a-life" in a sense similar to Singer's "persons" (Regan, 1983, pp. 243, 247). On the other hand, Descartes and neo-Cartesians like Carruthers view consciousness as an all-or-nothing trait, and deny that animals can be self-conscious.

DeGrazia distinguishes three kinds of self-awareness: bodily self-awareness (awareness of our bodies as distinct from other objects), social self-awareness (awareness of ourselves in relation to others in a social group), and introspective self-awareness (awareness of some of our own mental states). He argues that each type admits degrees. The last is the most complex one. His general standpoint is that

> our inevitable conclusion is that *self-awareness is not all-or-nothing but comes in degrees and in different forms*. This conclusion is important because it opposes a long tradition of speaking and theorizing about self-awareness as if it were all-or-nothing (DeGrazia, 1996, p. 182)

DeGrazia appears to mean that some animals have bodily self-awareness, others also social self-awareness, and some even introspective self-awareness (some apes).

DeGrazia also discusses language and moral agency. With regard to language, he mentions some older work on teaching apes to use sign language (DeGrazia, 1996, pp. 183–187). He also refers to, for example, the more recent work of Sue Savage-Rumbaugh on bonobos (pygmy chimpanzees) (DeGrazia 1996, pp. 187–198). This work took an interesting turn when Kanzi, a young chimpanzee, by observing his mother using a keyboard with signs learned much more quickly than his mother. He also learned to understand spoken English, which was used in Kanzi's presence and also directly to Kanzi (Savage-Rumbaugh, Shanker, Taylor, 1998). Pär Segerdahl, William Fields, and Sue Savage-Rumbaugh have recently discussed Kanzi's achievements in a book that stresses how Kanzi spontaneously learned the language in a culture shared with human beings (Segerdahl, Fields, Savage-Rumbaugh, 2006).

DeGrazia's conclusion is that language—like self-awareness—comes in different kinds and degrees. This holds true also of moral agency, which is another trait discussed by DeGrazia (DeGrazia, 1996, p. 204). Moral agency can be a matter of virtuous living but also of deliberation. DeGrazia refers, for example, to the work of Frans de Waal, who has shown that chimpanzees take reciprocity seriously (DeGrazia, 1996, p. 200; de Waal, 1982, pp. 175–177; see also Aureli and de Waal, 2000; Bekoff, 2005). It might be the case that some animals may exhibit virtuous living, but no non-human animal can manifest ethical deliberation (DeGrazia, 1996, pp. 203–204). This is how DeGrazia summarizes his view of moral agency:

> The fact that there are several defensible ways of understanding moral agency—which involve different capacities that are not all-or-nothing—suggests that this trait, like self-awareness and language, admits of both kinds and degrees (DeGrazia, 1996, p. 204).

This is not the place to take a definite stand with regard to self-awareness, language, and moral agency, but I agree with DeGrazia that if we accept that different types of each trait exist and also different degrees, then it is likely to be the case that some animals have some of these traits.

Self-awareness is ethically relevant. We have seen that both Singer and Regan stress our special obligations to animals that are self-aware. The best candidates for animals with self-awareness seem to be chimpanzees and gorillas. This provides strong reason for not carrying out experiments on these species, other than as the very last option (see below).

F. Critical Remarks

Let me end this section by making two critical points regarding DeGrazia's views. DeGrazia's conceptual analysis of sentience and consciousness is very enlightening. I think he is right when he argues:

> We know that we humans are conscious and that some of our mental states are typically potentially conscious. Given evolutionary continuity, neurological and behavioral analogues between humans and animals can ground attributions of similar mental states to them (DeGrazia, 1996, p. 103–104).

One word needs to be stressed, however, namely "can." We still do not know things for certain. More research into the minds of animals is crucial. In this research the differences between different species are especially vital to investigate.

My first critical point concerns DeGrazia's use of the theory of evolution. I agree with him that evolution suggests that animal sentience is biologically functional and that we have evolutionary reasons to believe that continuity exists between human beings and other related animal species. However, as we saw in the previous chapter, evolution cannot guarantee similarity. There might be crucial differences also between closely related species. DeGrazia runs the risk of overstating evolutionary continuity. We need to be very careful in attributing particular mental characteristics to other species.

Moreover, I find DeGrazia's view of animal welfare wanting. He comes close to what I have called a "feeling-based" account (DeGrazia, 1996, pp. 219–231). Above I have instead followed Fraser's suggestion that a comprehensive view of animal welfare is a better alternative. Such a view includes not only feelings but also biological functioning and natural living. However, I have also stressed that the feeling-based concern is the most important one in animal experimentation.

6. Animal Welfare in Animal Experimentation

After these discussions of animal welfare in general, let us turn to animal welfare aspects of animal experimentation. I have already made some comments on this in the context of the 3Rs. Now, I will discuss the animal welfare aspects of different stages of an animal experiment in a more systematic way. I choose to focus on the purpose of experiment, alternatives to the experiment, species and numbers, pre-procedural concerns, experimental procedures, post-procedural concerns, and endpoints (*cf.* Smith and Boyd, 1991). This brief presentation should be viewed as a kind of checklist regarding animal welfare in animal experimentation.

It is vital that these animal welfare aspects are considered in advance. One way of institutionalizing this can be seen in countries where researchers must submit an application for each particular project to an ethics committee of animal experimentation before the experiment is carried out. In the checklist below I will presuppose such a procedure.

A. Purpose

Purpose is extremely important. Which purposes are vital and which are not? We have already met different possible purposes in the Introduction. One important purpose is to obtain knowledge that is oriented toward improving health. It could be the health of human beings or the health of animals. Another purpose is to obtain basic biological or other knowledge.

With regard to human biomedical research, it is useful to make a distinction between direct and indirect purposes. By "direct purpose" I mean what the scientist directly intends to do, regardless of possible future applications. This direct purpose is related to a specific research question or set of questions, which the scientist attempts to answer through the experiment. An "indirect purpose" is an intended or expected future application. We may discern two major direct purposes: to obtain basic biological knowledge and to obtain health-directed knowledge such as knowledge about causes of disease or treatment. Often scientists having the direct purpose of obtaining basic knowledge justify their research by pointing out more or less probable future health-directed applications.

Agricultural research may have similar direct purposes: to obtain basic biological knowledge or to obtain health-directed knowledge. This latter knowledge concerns animal health. In addition, it may also aim at improving farm animal productivity.

From an ethical point of view, the purpose of a study is of vital importance. It restricts the range of possible methods, whether genetic or nongenetic. It is also closely related to expected benefit. The key problem is to determine which purposes have enough weight to justify an animal experiment. As we saw in the MORI poll described in Chapter One, many people assign much more weight to health-oriented research than to basic research (Aldhous et al., 1999). On the other hand, it is obvious from the history of science that many medical inventions would never have seen the light of day in the absence of basic research with no foreseeable applications. With this in mind, at least some basic research involving animal harm is justified, although it is difficult to draw the line. Some would argue that all or nearly all human research interests carry enough weight, others that severe—or even moderate—animal harm may outweigh some basic knowledge interests and sometimes even some health-oriented ones.

B. Alternatives

Another aspect concerns necessity. Are there any alternatives to using animals in order to attain the goal? In the discussion of the 3Rs, I mentioned some alternative methods such as using animals without ability to feel pain, conducting computer simulations, or carrying out experiments in cell culture. Such methods are important in research. Many scientists argue, however, that it often is necessary at some stage in the research process to use whole, intact, sentient animals. For example, we may start by testing the effects of a potential pharmaceutical in cell culture, but sooner or later we have to investigate how the drug functions in the body as a whole.

It is vital that scientists state their reasons for not using other methods, for instance by clarifying that alternatives simply do not exist in the case at hand.

C. Species and Number

The question about species is crucial. It is vital that researchers clarify what anatomical, physiological or other characteristics make them choose a particular species and strain given the scientific objectives of the project. As we saw in the presentation of public attitudes in Chapter One, many people find it ethically relevant whether mice or chimpanzees are used. The general rule in the scientific community should be to use animals as "low" in the animal series as possible.

The number of individual animals is also ethically relevant. We have already discussed this issue in relation to one of the 3Rs, namely reduction. We should reasonably not use a higher number of animals than necessary. I will return to this issue in the section on ethical balancing.

D. Pre-Procedural Concerns

The treatment of animals before the experimental procedure is carried out is also vital. Commonly, the animals are kept in the animal house of the university or other research institution and handled by experienced animal staff. Here it is crucial that the animals have some opportunity of natural living, and that their cages are environmentally enriched in order to stimulate positive affective states.

E. Experimental Procedures

Naturally, the welfare of the animals during experimental procedures is vital. As the third one of the 3Rs states, the experiments should be refined to ensure that suffering is minimized. The researcher should choose procedures that inflict the least amount of pain, distress or morbidity given the purpose of research. For example, the number of surgeries should be minimized, less invasive surgery should be carried out, and a less toxic adjuvant should be used.

All procedures—surgical and non-surgical—should be stated in advance. The dose, frequency, and method of administration of chemical agents should be specified. The method, volume, and frequency of blood and tissue sampling should be clarified. The time frame and endpoints of the experiment must be clearly defined.

The responsible scientist has also to consider in advance which measures to undertake to handle pain and distress. It can be a matter of providing anesthesia before surgical interventions or of giving post-surgical or other pain relief. In both cases, the dose and method of administration of any drugs should be stated. In the case of unexpected or unacceptable suffering it might be justified to euthanize the animal.

As indicated, feeling is the most important animal welfare concern in this context, but functioning is also important. It is vital to observe whether the animals exhibit clinical symptoms, that is, a reduced welfare as regards looks, function, and behavior. The clinical symptoms may be intended, as in the case of disease models, but they may also be unintended. In either case, the scientist should be prepared to handle the situation properly.

F. Post-Procedural Concerns

All procedures and induced conditions that will potentially cause more than minor pain, distress or morbidity should be identified, and the magnitude and duration of any adverse effects the animals may experience during the post-procedure period should be specified. The frequency and duration over which post-procedure monitoring of the animals will be performed should also be decided in advance. The appropriate intervals may be determined by the nature of the interventions, the degree of potential post-procedure pain, the likely duration of the pain and possible complications. For example, monitoring is often more important during the immediate post-surgical period, during the latter stages of tumor induction, and in toxicology experiments that have a high degree of morbidity.

A need exists for both feeling-based and function-based approaches to animal welfare. In assessing pain, the scientist should use behavioral and physiological parameters. Food and water intake, mobility, body temperature, and general healing of surgical incisions should be observed.

If it is expected that an animal may be subjected to more than minor pain or distress during the post-procedure period, analgesics should be provided prophylactically. In this case, the dose, method, frequency, and duration of administration of the analgesic agents should be specified.

G. Endpoints

Finally, we have the issue of endpoints. An "endpoint" is the point at which an animal shall be taken out of the experiment and be euthanized. What is an ethically acceptable endpoint? The criteria must be clearly stated. One condition can be that the purpose of the experiment is attained. Animals should be euthanized at the earliest possible endpoint given the scientific objectives. Another condition can be that the animal exhibits signs of unacceptable suffering due to, for example, tumor size, signs of significant pain, clinical symptoms or morbidity, and inability to feed. A third condition can be that the animal exhibits signs of unexpected suffering. There may be causes for premature euthanasia that are unrelated to the experimental procedures.

The method of euthanasia should be clarified and justified. It depends on the species, size of the animal, and its ability to quickly and painlessly produce a loss of consciousness and death. The scientists need to clarify what criteria will be used for determining that euthanized animals are dead. For example, they may use physiological parameters such as cessation of heartbeat and respiration for a particular period of time.

7. Ethical Balancing in Animal Experimentation

We have seen that ethical balancing is central to imaginative casuistry. We have also seen that it is central to the weak human priority view, not least in my version. In assessing an animal experiment, the expected human benefit and expected animal harm are to be balanced against each other. Some would argue that this is only a matter of balancing animal welfare and human welfare concerns. Others would argue that also animal concerns other than welfare concerns are to be included in the balancing, for example animal integrity. We have also seen that balancing is central to the comprehensive approach to animal welfare. Different animal welfare concerns may sometimes be in conflict with each other, making balancing necessary.

Moreover, we have found that the participants in the British MORI poll (see Chapter One) exhibited a quite sophisticated ability for case-by-case balancing, weighing potential human benefits against animal suffering and the species involved. This is interesting given the discussion of different ethical prototypes of animal experimentation above. Many people seem to come close to the weak human priority prototype with its balancing approach.

With all this in mind, it becomes vital to analyze what balancing is, what is to be balanced, and different methods of balancing.

A. The Nature of Balancing

Several problems arise with regard to the nature of balancing.

First, it is necessary to make a distinction between balancing before an animal experiment is carried out and balancing after the experiment has been conducted. In the former case balancing is part of pre-experimental decision-making, in the latter balancing is part of post-experimental assessment. These two situations give rise to different problems. In pre-experimental decision-making, we can predict fairly well the expected animal suffering, although unexpected things may happen. The expected human benefits are much more difficult to predict. In an assessment just a short time after the experiment, we know more exactly how the experiment has affected the animals. By means of extrapolation we also know more about potential benefits to human beings than before the experiment. However, the actual benefits to human beings are still unknown. It may take several years before we have this knowledge, and even then it might be very difficult to assess the precise contribution to this knowledge of this particular experiment. It is only a small part of a series of experiments leading to the new knowledge.

Second, we have the problem of aggregation. In classical utilitarianism, balancing is a matter of adding and reducing quantities of "utiles" such as pleasure and pain. This presupposes that aggregation is possible, that is, that the variables are commensurable. This presupposition is questioned by The Boyd Group, which stresses incommensurability, although they still believe that balancing is appropriate:

> It might be argued that weighing as suggested here is not possible, since there are no units of human (or animal) benefit and of cost to animals which could make these commensurable. Certainly, if weighing is thought of in terms of a mathematical calculus, this is correct. In every-day life, however, personal, professional and political judgements on moral issues normally require the weighing of factors and considerations which cannot be quantified with mathematical precision. A judge, for example, weighing a plea for mitigation of sentence in the "scales of justice" carries out a procedure of this kind (Smith and Boyd, 1991, p. 140).

As mentioned above, Ryder has argued that the pleasure and pain of individual animals cannot be aggregated:

I consider that it is the intensity and duration of pain of the individual that is most important. You cannot aggregate pain scores meaningfully across individuals. It is better, therefore, to inconvenience ten or a hundred animals than to cause severe pain to one. *So, the aim should be to reduce the pain felt by individuals, not the reduction in the total number of animals used* (Ryder, 1999, p. 40).

Ryder has a point (although I cannot accept his view that the number of animals feeling pain is not an ethically relevant aspect of its own: see above); we cannot literally aggregate the pleasure and pain of individual animals. A better understanding of the nature of balancing human benefit and animal harm is that it is a matter of making "trade-offs" (Ryder, 1999). A moral priority could be to try to reduce the pain of the maximum animal sufferer in each case, as Ryder (1999) suggests. Another moral priority could be to reduce the pain of future human sufferers, that is, patients. Balancing expected human benefit and animal harm is making a trade-off between these two aspects with regard to individuals, not aggregates.

Another problem concerns what is to be balanced. According to classical hedonism, it is pleasure and pain, but we have just seen Ryder's objection to that. Preference utilitarianism, on the other hand, focuses on preferences. The problem of aggregation becomes even more obvious in this case. Pleasure/pain and preferences are incommensurable. According to the comprehensive approach to animal welfare, different animal welfare concerns—functioning, feeling, natural living—are to be balanced. Fraser recognizes the problem and calls them explicitly "incommensurable variables." What about respect for integrity? Perhaps we should include violation of animal integrity as a cost in the balancing (*cf.* Stafleu *et al.*, 1999)? This would be to include a variable that is even more different compared to the welfare variables, and this would make the talk of balancing even more problematic. The conception of a trade-off is more appropriate than the conception of a balancing.

Fourth, we have the problem of the relative weight of the incommensurable variables. What is the relative weight of human welfare and animal welfare? As we have seen, the different prototypes of animal experimentation discussed in previous chapters give different answers. The strong human priority prototype would always give human research interests more weight than animal harm. The weak human priority prototype would commonly do so, but not always. The prototype of equal consideration of interests would not consider the aspect of who has a particular interest—whether human or animal—ethically relevant. In practice, this would imply that most research protocols would not be considered to carry enough weight.

A fifth problem concerns likelihood. A key issue in animal experimentation is whether the likely human benefit outweighs the likely suffering of the animals. What could "likelihood" mean in this context? The best way to interpret this concept is in terms of expected value. We can hardly know anything about "objective probabilities" regarding future applications of scientific knowledge, whatever that term may mean. It is better to understand likelihood in terms of a subjective degree of trust or expectancy. If the likelihood of human benefit of a particular animal experiment is low, its weight is perhaps not enough to justify the experiment. If it is high, it may contribute to the justification of the experiment. Assessing the likelihood of benefit of a particular experiment is very difficult. Commonly, the likelihood—in a subjective sense—of producing medical benefit is higher for health-oriented experiments than for basic ones. The reason for this is that they are aimed directly at contributing to solutions to health problems and are closer in time to such solutions than are basic experiments. This holds true despite the historical fact that basic research may sometimes have revolutionary implications and produce vast human benefit in the long run.

Finally, we face the problem of metaphor. Balancing is itself a metaphor, namely one of a balance with two scales. It can be interpreted in different ways by means of other metaphors. At least three such metaphors exist: aggregation, trade-off, and prioritizing. "Aggregation" is a metaphor of addition and reduction. It presupposes that the variables are commensurable. "Trade-off" is a metaphor of bargaining and does not presuppose that the variables are commensurable. "Prioritizing" is a metaphor of placing someone or something before someone or something else, that is, a matter of ordering or ranking. This metaphor does not presuppose commensurability, either. The fact that all the variables are incommensurable makes aggregation the least appropriate metaphor. Both trade-off and prioritizing are better in this regard. However, it is better to understand balancing in animal experimentation in terms of trade-off than in terms of prioritizing, because trade-off more clearly indicates that there may be opposite interests—human and animal—involved and that the goal is to reach a conclusion that somehow satisfies the opposing interests. Prioritizing is more neutral and does not recognize opposing interests.

B. Balancing in Practice

Let us take a brief look at different methods of carrying out balancing in practice.

Perhaps the most well known method is the already mentioned cost/benefit balancing in terms of utilitarian aggregation. According to this method, we may add and reduce "utiles" in a simple arithmetical way. A utile could be a unit of pleasure or well-being and its corresponding negative value could be pain or suffering. In assessing an animal experiment, we add the expected human benefit of, for example, 10 utiles and reduces the expected animal harm of, for example, –8 utiles. The aggregated result is 2 utiles.

A historically important example is Jeremy Bentham. In the following poetic form, he summarizes his method:

> Intense, long, certain, speedy, fruitful, pure—
> Such marks in pleasures and in pains endure.

Such pleasures seek if private be thy end:
If it be public, wide let them extend
Such pains avoid, whichever be thy view:
If pains must come, let them extend to few (Bentham, 1789, Chapter IV).

This method—called the "Felicific Calculus" or the "Utility Calculus"—could in principle establish whether a considered act is right or wrong. The variables of the pleasures and pains included in this calculation—which Bentham called "elements" or "dimensions"—were:

(1) intensity,
(2) duration,
(3) certainty or uncertainty,
(4) propinquity or remoteness,
(5) fecundity: the probability it has of being followed by sensations of the same kind,
(6) purity: the probability it has of not being followed by sensations of the opposite kind, and
(7) extent: the number of persons to whom it extends.

Bentham did not discuss animal experimentation, but we have already met his general view on the moral standing of animals: "The question is not, Can they *reason?* nor Can they *talk?* but, *Can they suffer?*" (Bentham, 1789, Chapter XVIII). Given this, it appears that Bentham would have thought it possible to apply the utility calculus also to animal experimentation.

Because the pleasure and pain of different individuals—whether human or animal—are incommensurable variables, this kind of calculus appears impossible in practice. For example, how should we balance mild suffering of many animals against severe suffering of a few? As Ryder points out, the pleasure and pain of individual human beings and animals cannot be aggregated. Aggregates cannot feel pleasure or pain, only individuals can (Ryder, 1999).

An alternative to utilitarian aggregation is Ryder's "trade-off" model. The trade-off model stresses that the relevant variables are incommensurable. But how should the trade-off be carried out if the variables are incommensurable? It appears that we are left with an intuitive method. This appears, on the other hand, too subjective to be entirely satisfying.

Perhaps it is possible to reach an intersubjective agreement on certain steps of procedure even if it is not possible on particular judgments? Several possibilities exist.

One option is a simple matrix model, for example a matrix consisting of two dimensions: the degree of severity of the experiment as regards animal suffering (mild, moderate, severe) is balanced with the expected benefit of the experiment (low, medium, high).

A slightly more complex model is "Bateson's cube" with three dimensions: animal suffering, probability of benefit, and quality of research. Each dimension has three possible values: low, medium, and high (Bateson, 1986).

The Boyd Group has developed Bateson's cube model in more detail. The assessment of the potential and likely benefit of the project consists of three steps. The first concerns the potential benefits of the project: social value, scientific value, economic value, educational value, other value (not all of these may be relevant in a given project), originality, timeliness, pervasiveness, and applicability. The second step consists of the assessment of the proposed approach: scientific merit, necessity and validity of the procedures, and quality of the workers and facilities. The third step is an overall assessment of likely benefits. The assessment of the likely cost to animals also consists of three steps. The first is an examination of the quality of facilities and project workers. The second is examination of the severity of the effects of husbandry and procedures on animals: type of animal used, husbandry and housing conditions, likely severity of adverse effects, provision of amelioration of adverse effects on animals, and number of animals. The third step is an overall assessment of costs likely to be imposed on animals. The Boyd group explicitly states that it makes no attempt to provide universally applicable rules for weighing costs and benefits (Smith and Boyd, 1991, pp. 138–147).

An attempt to go further and use a numerical method in balancing has been suggested by Stafleu *et al*. Several relevant parameters are assigned different numerical values, which are put into formulas. Stafleu *et al*. argue that the different aspects of expected human benefit is commonly not included in a sufficiently detailed way. Their numerical model consists of eight steps:

(1) description of the ultimate aim of the experiment,
(2) determination of the weight of the human interest,
(3) computation of the total interest score of the ultimate aim,
(4) assessment of the relevance of the experiment,
(5) calculation of the interest of the experiment for human beings,
(6) assessment and scoring of the harm to the interest of animals,
(7) computation of the harm score for animals, and
(8) assessment of the ethical acceptability of the experiment (Stafleu *et al*., 1999).

Of special interest is the suggestion by Stafleu *et al*. that violation of integrity be included as a cost (Stafleu *et al*., 1999; *cf*. Heeger, 1997; see Chapter Three). They assign this cost a particular numerical value. Possible criticisms are that the value is too arbitrary or that many would assign a much higher value—perhaps even a value as high that would in effect rule out almost every animal experiment.

C. Proposal: A Matrix for Ethical Trade-Off

The focus of this book is on the social ethics of animal experimentation, not the private or personal ethics. The problem of ethical balancing is central for research teams, ethics committees on animal experimentation, and agencies. This means that what is important is dialogical balancing rather than individual balancing. The balancing is to be carried out in dialogue within social groups and in dialogue between social groups. An appropriate model of balancing must take this social aspect into consideration. With this in mind, I propose the following method of balancing.

As suggested above, balancing should be understood in terms of trade-off instead of aggregation. Trade-off is more appropriate, taking into consideration that what is to be balanced are incommensurable variables.

A simple matrix model is preferable to more complex models and advanced numerical models. It is easier to handle in social groups such as research teams and ethics committees on animal experimentation. It is more transparent, making it more open to public accounting. The matrix has two basic dimensions: expected level of human benefit (low, medium, high) and expected level of animal harm (mild, moderate, severe). In this latter dimension may be included not only suffering but also violation of integrity. An additional dimension concerns species.

An experiment with low expected human benefit would be one aiming at basic knowledge with no obvious health applications. An experiment with medium expected human benefit would be one aiming at understanding causes of and finding treatments for non-life-threatening diseases. An experiment with expected high human benefit would be one aiming at understanding causes of and finding treatments for life-threatening diseases.

Animal harm consists of two parts: low animal welfare and violation of animal integrity. Animal welfare is to be understood in terms of the comprehensive approach to animal welfare, that is, as encompassing concerns of functioning, feeling, and natural living. These concerns are to be ordered, taking the experimental context into account. As argued above, animal welfare can in this context be expected to focus primarily on avoidance of negative affective states but also—in housing and care—on promotion of positive affective states by provision of enriched environments. In the context of using animals as disease models and for gene knock-out experiments, functioning well (including health) can be expected to have low priority. Natural living can be expected to have low priority, because of the necessity to keep the animals in cages for reasons of scientific control.

Violation of animal integrity should be included as a cost in the balancing (*cf.* Stafleu *et al.*, 1999; see Chapter Three). Animal integrity places the burden of proof on those who intend to carry out animal experimentation.

An experiment with severe expected animal harm would be one with severe suffering. An experiment with moderate expected animal harm would be one with moderate suffering. An experiment with mild expected animal harm would be one with no suffering or even improved welfare.

As an illustration of how to use the matrix, let me present two prototypical options at the extremes. One experiment may have low expected human benefit because it aims at basic knowledge with no obvious health application and severe expected animal harm because it will lead to severe animal suffering. According to the "weak human priority" view, this experiment should probably not be carried out. Another experiment may have high expected human benefit because it aims at medical treatment of a life-threatening disease and mild expected animal harm because it leads to very limited suffering. This experiment is much easier to accept from the point of view of "weak human priority." In between these two extreme prototypical cases, there may be difficult non-prototypical cases that require serious ethical deliberation and trade-off.

Let me also illustrate how considerations regarding different animal species might be included in the use of the matrix. Take mice and chimpanzees. If mice are used the present approach would be much more permissive than if chimpanzees are used. Experiments on mice would be allowed in all cases except when the expected suffering is severe and the expected human benefit is low. Experiments on chimpanzees on the other hand would be accepted only if the expected suffering is extremely limited and the expected human benefit is extremely high (this may perhaps never be the case in practice).

How should this view of balancing be characterized from the point of view of ethical theory? In Chapter Three I stated that imaginative casuistry—my preferred general ethical approach—suggests a plurality of values and a mixture of consequentialist and deontological considerations. In the ethics of animal experimentation my view is best described as mainly consequentialist but with certain deontological restrictions.

The view is consequentialist because of its focus on human benefit and animal harm, and because this benefit and this harm are to be balanced in particular cases. Central to human benefit is the reduction of human suffering and disease. Central to animal harm is animal pain and suffering. However, the focus is not only on subjective experiences—suffering, pleasure, and so forth—but also on objective states such as functioning. This means that this consequentialist approach is not purely utilitarian in this respect, although it is true that in animal experimentation pain and suffering are the most important ethical concerns.

Neither is it purely utilitarian with regard to balancing, because it does not accept aggregation of utiles. Numbers are ethically relevant but the pleasure and pain of individual human beings and animals cannot be aggregated. The pleasure and pain of different individuals—whether human or animal—are incommensurable variables. Therefore, trade-off instead of utilitarian aggregation is preferable.

Also in another respect my proposal is not purely utilitarian regarding balancing, namely that the leading principle is not equal consideration of interests but a weak priority of human interests over animal interests. This human priority is due to social bonding, which is expressed in care for our children and also other human beings. The priority is weak in the sense that human interests are not outweighing animal harm in all cases. This weak priority given to human beings can be viewed as a kind of deontological restriction on the consideration of interests. Interests are not to be considered equally but to some extent unequally.

I suggest also other deontological restrictions. As we have seen, one restriction is that if the expected human benefit of an animal experiment is very low and the expected animal pain and suffering is severe, the experiment should not be carried out.

Another deontological restriction is that if animal integrity is severely violated during the experiment (painless euthanasia not included), it should not be carried out. This holds true even if the animals do not feel pain or suffer. Animal integrity might be violated not only if the animal suffers severely but also if it does not function well or if it is leading a too unnatural life. Animal integrity in this sense is not a matter of all-or-nothing but a matter of degree.

Finally, I propose a deontological restriction regarding species. As we have seen, my view is more permissive for some species than for others.

None of these deontological restrictions to consequentialist considerations is strictly fixed. A judgment has to be made in each particular case.

D. Additional Proposals: Precedents and Feedback

Let me end this chapter with two additional proposals.

Measures must be undertaken in order to reduce subjectivity and arbitrariness in making the ethical trade-off. For example, in determining the likelihood of future human benefit, optimists and pessimists might make different judgments. One method of counteracting this risk is analogical reasoning based on precedents. This is particularly crucial in animal ethics committees and agencies. The way an ethical trade-off has been made in a similar previous case should be taken into account in the ethical trade-off in the new case. This analogical reasoning is typical of casuistry (Jonsen and Toulmin, 1988).

There should also be a procedural feedback system. Earlier successes or failures of the research program should be taken into account in the assessment of individual projects. In this way, the assessments of the animal ethics committees or agencies may be carried out in a more impartial way.

Six

GENETICALLY MODIFIED ANIMALS IN RESEARCH

1. Implications of the Five Prototypes

Let us apply the five ethical prototypes of animal experimentation—presented in Chapter Two—to genetically modified animals, including cloned animals.

The human dominion prototype is very permissive regarding genetic modification of animals. We have no direct duties to animals, only indirect ones. Whether it is ethically acceptable to genetically modify animals depends solely on whether it is in line with moral responsibility to human beings. Human benefit is what counts.

According to the strong human priority prototype, scientists may produce and use genetically modified animals in response to human moral right or human need. Cohen stresses explicitly the benefit of transgenic and knockout mice in research (Cohen in Cohen and Regan, 2001, p. 78). Genetically modified animals may be useful in the search for basic biological knowledge and new treatments of human disease. Scientists should try to minimize the suffering of genetically modified animals, but animal suffering cannot be a reason not to carry out an experiment for which good scientific reasons exist.

The weak human priority prototype is more restrictive, although more or less radical versions have been proposed. This model stresses that the ethical acceptability of animal experiments has to be established on a case-by-case basis. It is acceptable to genetically modify animals only if we have strong reasons and only if animal suffering is minimized. The suffering of genetically modified animals may sometimes outweigh human benefit. Midgley appears quite critical toward many cases of experiments involving genetically modified animals. She stresses that the negative gut feeling many people have with regard to genetic modification should be taken seriously (Midgley, 2000). Below I will analyze her argument in more detail. Let me just point out that it is possible for an adherent of weak human priority to be more positive. This is the view on genetic modification of animals that I will defend.

The prototype of equal consideration of interests is even more restrictive with regard to genetic modification. According to this view, we should not genetically modify animals unless we have extremely strong reasons. The suffering of genetically modified animals generally outweighs human benefit. However, genetic modification is not ruled out in principle. For instance, producing human proteins in sheep milk could be possible without animal suffering.

The animal rights prototype, on the other hand, rules out genetic modification in principle. It is ethically unacceptable to genetically modify animals, since this would violate their inherent value. Because all animal experimentation is wrong, experimentation involving genetic modification is wrong.

Thus, the human dominion prototype and the animal rights prototype give definite views on the ethical acceptability of genetic modification of animals in research. The former is completely permissive, while the latter is entirely prohibitive. The other three prototypes only give some general direction. They call for a more focused discussion of pros and cons. With this in mind, I will investigate the concerns raised by genetic modification of animals in research in more detail. Basically, three main concerns have been expressed in the debate: scientific concerns, intrinsic ethical concerns, and animal welfare concerns.

2. Scientific Concerns

I will focus on three different scientific concerns, namely the purpose of experiments involving genetically modified animals, the design of such experiments, and, finally, the reasons for using genetic modification methods instead of non-genetic ones.

A. The Purpose of Experiments Involving Genetically Modified Animals

Scientists from a great variety of fields produce and use genetically modified animals in research. Examples of such fields are genetics, cell biology, molecular biology, physiology, immunology, cancer research, neuroscience, pharmacology, diabetes research, and cardiovascular research. Also in the agricultural sciences researchers are involved in this kind of research. Genetically modified animals may be produced as tools for unspecified uses in research, for example by a transgenic facility serving an entire university, or for a specified purpose.

In human biomedical research and in agricultural research, we discern two major purposes for producing or using genetically modified animals: to obtain basic biological knowledge and to obtain health-directed knowledge. In agricultural research, there may also be an additional purpose, namely improving farm animal productivity.

Genetically modified animals are often used as disease models, that is, as models for understanding disease processes and causes of disease, and for testing new pharmaceuticals and other therapies (Nuffield Council on Bioethics, 2005, p. 127). Many scientists consider genetically modified animals to be better models than conventional ones for many purposes.

Genetically modified animals are also used for obtaining knowledge of gene function. The Human Genome Project is now completed. We have knowledge of the entire human DNA sequence. The next step is to understand the function of genes and how they are regulated. In this research, genetically modified animals play a key role. By "knocking out" or "overexpressing" single genes, their function can be detected. This can be done in animals, but—for ethical reasons—hardly in human beings.

Another option is toxicity testing. Genetically modified animals are used in testing chemicals and drugs to ensure that they do not cause cancer. A modified gene is inserted, making the development of cancer occur much earlier than normal. Genetically modified mice are also used for testing effects that previously could only be tested on primates (Royal Society, 2001).

Genetically modified animals may also be used as bioreactors. They are engineered to produce human therapeutic proteins in their milk. Examples are human blood-clotting factor IX, alfa1-antitrypsin, insulin, and human growth hormone.

One further therapeutic use is xenotransplantation. The reason behind xenotransplantation is to increase the number of organs available for human transplantation by using animal organs. Such organs give rise to immunological reactions. Animals are genetically modified to prevent the expression of proteins that cause immunological reactions. The main interest is to obtain pig hearts for human transplants.

It is also possible to use genetically modified animals in agriculture. Farm animals may be genetically modified in order to improve the immune system, to improve the resistance against parasites, or to increase productivity by making them grow faster. The composition of meat and milk may be changed. The composition of wool may be modified in sheep. While no livestock to date has been genetically modified for food, fish—for example, salmon—has been modified.

If we turn to animal cloning, we also see that with regard to this technology there can be many different reasons for carrying out the experiments. In several cases, genetic modification in the ordinary sense is combined with cloning. The reason is that when a genetic modification has been successfully achieved, which is not easy, then the result can be preserved by cloning the genetically modified animal. This is potentially the case with xenotransplantation and bioreactors, but also with breeding animals such as farm animals or sports animals with special traits (although cloning is probably not good for breeding generally and in the long term).

Animal cloning may also be an important step on the way to human cloning, although human reproductive cloning is commonly not an objective for carrying out animal cloning experiments. Many scientists conducting this kind of research explicitly reject human reproductive cloning. Human therapeutic cloning might, however, be an objective. This type of cloning is a way of treating diseases. It is a special form of stem cell therapy, in which the embryonic stem cells are derived from the patient himself or herself rather than from an embryonic stem cell line originating from another person. The advantage of therapeutic cloning over ordinary embryonic stem cell therapy is that immunological rejection can be avoided (Hochedlinger and Jaenisch, 2003).

Another purpose is to obtain basic biological knowledge. Although much of this kind of research can be called "applied," it is obvious that by struggling with different technological problems of achieving the desired cloning, new fundamental knowledge of biological mechanisms and processes is gained. This aspect was emphasized by the researchers that produced the first cloned dog (Lee *et al.*, 2005).

Cloned animals may also be of special value in animal experimentation. The reason is that in experiments as similar individuals as possible are desirable. Cloned animals are (almost) identical, and strengthen the reliability of the experiments. Uncontrolled differences do not influence the results.

Two other applications of animal cloning are directly concerned with animals as such. The first is saving endangered species. If only very few individuals are still alive of a particular species, they might be preserved by means of cloning. The more basic reason for this can be preservation of biological multiplicity. This is a central principle in environmental ethics. Another application is replacement of deceased companion animals, although this is probably an option only for the very rich.

An ethical question that arises with regard to both genetic modification of animals (in the ordinary sense) and animal cloning is whether all the purposes for which these technologies can be used are ethically acceptable. The weak human priority view that I propose would consider some of the purposes more justifiable than others. Genetic modifications that are disease-related would be acceptable, while the cloning of pets would be much more doubtful.

B. Methods of Genetic Modification

Another scientific concern is the design of the experiments. Several different methods exist for producing genetically modified animals. Let us take a closer look at the two main methods: pronuclear microinjection and the embryonic stem cell method.

In presenting these methods, I choose the mouse as an example. It is by far the most common species that is genetically modified, and both main methods for genetic modification are applicable to them. With regard to rats, primates, and farm animals, the embryonic stem cell method has to date met with only limited success. Pronuclear microinjection is the most common method for these species (Royal Society, 2001, p. 6).

Pronuclear microinjection is the classical method of transferring genes. The genes may come from the same species or from another one. The use of this method was first reported in 1980 (Gordon *et al.*, 1980). Applied to the mouse, the main steps are as follows.

(1) *Superovulation.* A female mouse gets hormone stimulation and releases 20–50 eggs. The common number is 8–10 eggs. Approximately 20 females are required to generate three to four genetically modified founders.
(2) *Mating.* The female mouse mates with a selected male during nighttime. A sign of successful mating is the occurrence of a vaginal plug.

(3) *Euthanasia and harvesting of eggs.* The donor female is euthanized (commonly by dislocation of the neck) and the fertilized eggs are harvested.
(4) *Microinjection.* With a very thin needle of glass a solution containing the foreign gene is injected into one of the two pronuclei of the fertilized egg (one pronucleus originates from the male and one from the female).
(5) *Surgical implantation* of the eggs into a pseudopregnant surrogate mother. Another female (the surrogate mother) is first made pseudopregnant by mating a sterile male (vasectomized or genetically sterile). The mating leads in mice to hormonal changes making it possible for the female to receive the fertilized eggs. The female is anesthetized and the embryos are transferred into the reproductive tract through a surgical incision. The skin is closed with skin clips, surgical staples or suture. The recipient female is then allowed to wake up.
(6) *Breeding.* When the pups are born and grown up, they are used for breeding. Three different generations emerge. Generation 0 (F0) is the founder generation—hemizygous for the new gene. It may consist of full transgenics—with the new gene in all cells—and mosaics—with the new gene present only in some cells (and some of these may be germ cells). By breeding a hemizygous full transgenic with a wild-type individual, half the generation 1 (F1) will inherit the new gene. Finally, by selecting F1 hemizygous brother-sister breeding pairs, a homozygous transgenic generation 2 (F2) is produced. Thereby, a genetically modified strain is generated. Some animals generated during the process of production are useless for the program of genetic modification. These may include non-transgenics from the different generations, germ-line and non-germline mosaics, F1 hemizygotes, and non-expressing transgenic animals.
(7) *Genotyping.* Animals belonging to the different generations are consecutively genotyped, in order to see which individuals have the inserted gene. In this regard, Southern blot analysis or polymerase chain reaction (PCR) are used. Tissue samples or blood is obtained by tail biopsy or ear notching (Polites and Pinkert, 2002; Houdebine, 2003; BVAAWF *et al.*, 2003, pp. S1:3–4).

A problem with pronuclear microinjection is random integration. The gene may end up in the wrong place, leading to unpredicted effects or no detectable effects at all. These effects may be reduced by the use of insulator or intronic sequences.

The next technique is the embryonic stem cell method. This method makes possible targeted genetic modification (Capecchi, 1989). This means that the integration is not random but controlled; the transferred gene ends up in exactly the correct place. The method can be used for insertion of a new gene or inactivation of a gene. The former is called "knock-in," the latter "knock-out."

The methodological steps are to some extent the same as in pronuclear microinjection, but completely new ones are also added.

(1) *Superovulation* (although this is less commonly used to increase the number of blastocysts).
(2) *Mating*.
(3) *Euthanasia and harvesting of eggs*.
(4) *Isolation and cultivation of embryonic stem cells*. Embryonic stem cells are derived from inner cell mass of early embryos, that is, blastocysts.
(5) *Electroporation in cell culture* in order to achieve homologous recombination. Embryonic stem cells are put into a liquid in a test tube or petri dish. Copies of the gene move freely among the cells. An electric current is sent through the liquid. This current forces the embryonic stem cells to open their cell membranes. In this way, the foreign gene can be integrated into the genome of the embryonic cells.
(6) *Blastocyst injection*. Embryonic stem cells with the desired genetic modification are injected into blastocysts.
(7) *Surgical implantation* of the embryos into a pseudopregnant surrogate mother.
(8) *Breeding*. Also with this method, three different generations emerge. In F0, some mice animals will be chimeric—containing both modified and unmodified cells—while the rest will be unmodified. Some chimeras will be germ-line chimeras. These are mated with wild-type individuals with the result that some of the F1 animals will be hemizygous for the new gene. By crossing these hemizygotes, fully transgenic F2 homozygotes are generated. As with pronuclear microinjection, many animals produced with this method are useless. Mice in F0 and F1 with embryonic stem cell contribution are commonly identified by coat color changes.
(9) *Genotyping* (Doetschman, 2002; Houdebine, 2003; BVAAWF *et al.*, 2003, pp. S1:4–5).

Let me also mention a few other methods. Conditional methods are becoming increasingly important (Rucker *et al.*, 2002; BVAAWF *et al.*, 2003, pp. S1:6–8; Houdebine, 2003). These methods make it possible to induce modifications that are tissue-specific or temporally specific. Genes can be activated or inactivated in the type of tissue we desire and at the point of time we wish.

A key example is the Cre-LoxP method. In this method different recombinations—deletion, translocation, inversion, insertion—are carried out by an enzyme from the bacteriophage P1 called Cre (*causes re*combination) recombinase. This recombinase recognizes short DNA sequences called LoxP (*lo*cus of *x*-ing over) on each side of the target gene (the gene is floxed = *f*lanked by *lox* sites). Generally, Cre-expressing mice are produced by pronuclear microinjection, while floxed mice are generated by the embryonic stem cell method. When mice of these types are crossed, the desired recombination occurs in the specified tissue. Flp/frt recombination systems have similar abilities. Another type of conditional method is inducible transgenes, for example, systems dependent on tetracycline or its derivates. Tetracycline is administered by injection or by adding it to the drinking water, and allows a transgene to be expressed only in a specific cell type and only when the tetracycline is given to the animals. For example, if a gene is removed, it may be replaced with a tetracycline-regulated transgene (Rucker *et al.*, 2002; BVAAWF *et al.*, 2003, pp. S1:6–8; Houdebine, 2003; Aiba and Nakao, 2007).

Another method for producing genetically modified animals is the retroviral method in which retroviruses are used as vectors for the DNA, but I will not present this method in detail here (Kim, 2002).

Finally, nuclear transfer technologies have been developed in which a nucleus from a somatic cell is transferred into an enucleated egg (Wilmut *et al.*, 1997). The nucleus can have modified or unmodified DNA (Tsunoda and Kato, 2002; Paterson *et al.*, 2002). In this latter case, the whole nuclear genome can be viewed as modified, since it is transferred from another animal into the egg. This somatic cell nuclear transfer is often called "animal cloning." It is vital to distinguish two types of animal cloning, reproductive cloning and therapeutic cloning (*cf.* the presentation of human cloning above).

The methodological steps in reproductive cloning are as follows:

(1) *Collection of somatic cells from the animal that is to be cloned.*
(2) *Collection of donated eggs.*
(3) *Removal of the nucleus from a somatic cell.*
(4) *Removal of the nucleus from a donated egg.*
(5) *Transfer of the nucleus of the somatic cell into the enucleated egg.*
(6) *Activation of the cloned embryo.* By means of, for example, electric impulses the clone is now reprogrammed from the adult state to the gamete or embryo state.
(7) *Surgical implantation of the embryo into a surrogate mother.*
(8) *Support to the surrogate mother to carry to term and give birth to the clone* (Tsunoda and Kato, 2002; Paterson *et al.*, 2002).

In therapeutic cloning the first six steps are the same as in reproductive cloning. Then the steps are as follows:

(7) *Derivation of embryonic stem cells.* At the blastocyst stage (around 5 days after conception), the embryonic stem cells are sucked out and the embryo dies.
(8) *In vitro culture.* The embryonic stem cells are cultured.

(9) *Differentiation.* The embryonic stem cells are stimulated to develop into the desired type of cells, for example, nerve cells, muscle cells, or liver cells (Hochedlinger and Jaenisch, 2003).

C. Providing Specific Reasons for Producing and Using Genetically Modified Animals in Research

In arguing for one particular method of genetic modification instead of another or instead of a non-genetic modification method, it is vital that scientists specify their reasons clearly. In an empirical study of applications submitted to ethics committees on animal experimentation in Sweden regarding production and use of genetically modified animals—carried out by my colleague Helena Röcklinsberg and myself—we found that this was often not the case. The reasons were often unclear and poor, and sometimes lacking (Nordgren and Röcklinsberg, 2005). If scientists wish to retain public confidence—in particular regarding the use of genetic modification methods rather than non-genetic ones—it is crucial that they present their reasons in a straightforward manner.

The general structure of these arguments may follow three patterns. The first pattern of reasoning is that, given a particular purpose, it is scientifically *necessary* to produce (or use) a particular genetically modified animal. The second pattern is that given a particular purpose, it is scientifically *better* to produce (or use) a particular genetically modified animal than to use other genetic modification methods or non-genetic ones. The third pattern is that given a particular purpose, it is scientifically *as good* to produce (or use) a particular genetically modified animal as to use other genetic modification methods or non-genetic ones (Nordgren and Röcklinsberg, 2005).

In order for the argument to work, two steps of justification are needed. First, the purpose must be justified. A good reason must be provided for believing that the purpose of obtaining this particular piece of basic knowledge or this particular piece of health-directed knowledge would be valuable to society. Second, the scientific necessity or suitability must be justified. A good reason must be given for believing that it is "scientifically necessary," "better," or "as good" to produce or use this particular genetically modified animal in order to realize this particular purpose (Nordgren and Röcklinsberg, 2005).

Let me give three examples focusing on the distinction between genetic and non-genetic modification methods in biomedical research. In the first, the purpose is to discover the function of a particular gene. In this case, the scientist could argue that it would be scientifically necessary to make a knock-out. No non-genetic modification methods are possible. In the second example, the purpose is to create an animal model of a particular disease, which is also possible by using non-genetic modification methods. In this case, the scientist wants to show that it would be scientifically better to make a genetically modified model than to induce the disease—for example, diabetes—using non-genetic modification methods. The third example is similar to the second, but in this case the argument is that the genetically modified disease model is as good as the non-genetically modified model (Nordgren and Röcklinsberg, 2005).

A key ethical question is whether non-genetic modification methods are preferable to genetic modification methods. Some may argue that if it is neither scientifically necessary nor better to produce (or use) genetically modified animals in order to achieve the purpose of study, but it is only as good as using non-genetic modification methods, then special ethical justification is needed. If the genetic modification method is only as good as the non-genetic modification method—which probably occurs only rarely in practice—they may argue that the non-genetic modification method is ethically preferable. A reason for this could be the unintended and unpredictable welfare effects that may occur in the production of genetically modified animals. To this it could be objected that there may be unpredictable welfare effects in all animal experimentation and that genetic modification methods may often make the outcome more predictable, for example when compared to selective breeding. Moreover, unpredictability is an issue only in the development of new lines, not in the use of existing ones. Another reason for preferring non-genetic modification methods could be that using genetic modification methods is probably much more difficult and expensive. The efforts and resources should be used as effectively as possible. A possible objection is simply that, in practice, no one would use a more difficult or expensive method than is necessary. If so, the argument carries no real weight. Another possible objection is that it would be immoral not to spend the money on research that we have good reason to carry out. A third reason for preferring non-genetic modification methods could be that genetic modification constitutes a more serious violation of animal integrity (see below). It could be objected, however, that surgery or pharmacological treatment may sometimes violate integrity even more.

The issue of alternatives to genetic modification methods should be put into a broader perspective. It is vital for scientists to retain public trust. If the general public finds genetic modification methods only as good as non-genetic modification methods, they will probably not accept them. This is a general consideration regarding technology. If a particular technology is considered necessary, the general public will probably accept it, even if the risks are high. If it is not viewed as crucial, people will probably not accept it. We see this very clearly with regard to genetically modified food. As several Eurobarometer surveys show, many people simply do not see the point with it and consequently they do not accept the risks. Regarding medical technology, they accept risks more easily, since this technology is viewed as crucial (Eurobarometer 58.0, 2003; Bonny, 2003). This means that it is vital for scientists to show very clearly that genetic modification methods are necessary or at least better than non-genetic modification methods for some experimental purposes.

3. Intrinsic Ethical Concerns

We find two types of intrinsic ethical concerns. The first regards all animal use. We have already met this concern in the discussion of the animal rights model. Recognition of the intrinsic (or inherent) value of animals raises the issue whether it is ethically acceptable for human beings to use animals at all. Some would argue that it is never acceptable to use animals as tools. Others would argue that it is not acceptable to use animals merely as tools, but that it can be acceptable to use them as tools as long as their intrinsic value is respected.

The second type of intrinsic concern regards genetic modification as such. Before any ethical balancing of advantages and disadvantages of genetic modification of animals is carried out, we must determine whether genetic modification is ethically acceptable in itself.

The possible intrinsic ethical concerns are constantly underestimated and neglected by scientists. Many scientists seem to reason only in terms of benefits and risks. Often intrinsic concerns are dismissed as being merely a matter of feelings. However, they are of key importance to the general public. Many people question genetic modification of animals on moral grounds. They view such modification as morally wrong in itself. The results of the Eurobarometer and the results of Macnaghten's focus group study show that people care about moral aspects such as violation of natural order (see Chapter One).

Intrinsic concerns are not merely a matter of feelings but also of thought. As we saw in our analysis of the weak human priority prototype, Midgley points out that reason and feeling are complementary aspects of the moral process. Feelings incorporate thoughts, and reasons are developed in response to feelings. She states:

> We find our way in the world partly by means of the discriminatory power of our emotions. The gut sense that something is repugnant or unsavory—the sort of feeling that many now have about various forms of biotechnology—sometimes turns out to be rooted in articulable and legitimate objections, which with time can be spelled out, weighed, and either endorsed or dismissed. But we ought not dismiss the emotional response at the outset as "mere feeling" (Midgley, 2000, p. 7).

Midgley's view requires some comment. I agree that we should take our feelings toward genetic modification of animals seriously, and I also agree that these feelings can be spelled out in terms of rational arguments. However, I think that Midgley does not distinguish clearly enough between different types of feeling. We have already met her view on the moral relevance of feelings of social bonding. Taking these feelings seriously and giving a weak priority to human research interests compared to animal interests is not prejudice but a moral imperative. These biologically grounded feelings of social bonding must be distinguished from "the yuk-factor" Midgley talks about in relation to biotechnology (Midgley, 2000, p. 7).

The central importance of "moral imagination" further underlines the supplementary nature of reason and feeling. Moral imagination involves envisioning intellectually alternative perspectives and arguments and emotionally empathizing with others. An important tool in moral imagination is the use of metaphor. Metaphors govern our thinking. The key question is: which metaphors should govern our thinking concerning genetic modification of animals? In this section, I will analyze and discuss some metaphors used in spelling out the intrinsic concerns related to genetic modification of animals.

The intrinsic concerns regarding genetically modified animals are caught in four different types of argument. The first is put in terms of "playing God," the second in terms of "violation of the natural order," the third in terms of "violation of animal *telos*," the fourth in terms of "violation of animal integrity." The first two arguments are general and concern all genetic modification. The last two concern only genetic modification of animals. As we will see, each type of argument appears in categorical and non-categorical versions.

All these arguments use different metaphors. The different versions of the arguments in addition use other metaphors.

A. "Playing God"

The "playing God" argument can be used as a general argument against all genetic modification, whether of animals, plants, or microorganisms. Midgley, commenting on the term, points out that

> *playing God* ... is actually a quite exact term for the sort of claim to omniscience and omnipotence on these matters that is being put forward (2000, p. 14).

In this way, Midgley urges us to take the "playing God" argument seriously. But what does the term mean more precisely? Basically, two different types of interpretation have been proposed, a religious and a secular. The religious version can be put forward from within different religious frameworks. I will focus here on the Judeo-Christian tradition.

In the US, the President's Commission report on *Splicing Life* concluded that "playing God" had no special religious meaning (President's Commission, 1982, p. 54) and translated the term into a secular concern about the consequences of exercising great human powers (Lebacqz, 1984, p. 33). Allen Verhey, on the other hand, argues that the term does have a special religious meaning, or rather, several meanings (Verhey, 2002). Ted Peters points out that it is a phrase that is foreign to theologians and is not common in a theological glossary, but that some religious spokespersons use the idea in discussions of genetics (Peters, 1997, p. 12). Some uses of the term have a negative connotation, others a positive one.

The religious "playing God" arguments against genetic modification are based on a negative religious interpretation of the term "playing God." To play God is to do what only God is allowed to do.

According to the positive religious interpretation of the term "playing God," human beings are to be the stewards of Divine creation. Joseph Fletcher uses the term in this sense when invoking us: "Let's play God" (Fletcher, 1974, p. 126).

Paul Ramsey acknowledges both interpretations in stating that although we are usually warned against "playing God," we are sometimes encouraged to "'play God' in the correct way" (Ramsey, 1970, p. 256). We are to "imitate" God (Ramsey, 1970, p. 259), like a child "playing" a parent.

Thus, the first religious objection to the "playing God" argument is that it is possible to talk about "playing God" also in a positive sense. We could be viewed as co-creators with God in carrying out genetic modification. It depends on the purpose of genetic modification whether it is ethically acceptable, that is, whether it is truly a matter of co-creating with God. This has to be determined from case to case. This view implies a non-categorical interpretation of the "playing God" argument against genetic modification. Only some cases of genetic modification are a matter of "playing God" in a negative sense.

This, in turn, suggests a second religious objection, namely that categorically condemning all genetic modification does not permit discriminating judgments (*cf.* Verhey, 2002). This means that the "playing God" argument in a categorical sense is not very helpful in practice.

Finally, we find the non-religious objection that the religious version of the "playing God" argument is not convincing for non-believers and therefore not useful in working out a governmental policy or a social ethic for a pluralistic society.

A secular version of the "playing God" argument has also been proposed. It is "playing God" in this sense that Midgley and the President's Commission appeal to. The key idea is that genetic modification is an expression of a hubris that will be punished in the end. Midgley talks about "the hype, the scale of the proposed project, the weight of the economic forces backing it" (Midgley, 2000, p. 8). The secular "playing God" argument is a criticism of the human attempt to control the non-human world by means of gene technology. Sometimes the ancient myth of Prometheus is referred to as a discouraging example. Prometheus stole fire from the gods and was punished.

As was the case regarding the religious interpretation, we find categorical and non-categorical versions of the secular interpretation too. An objection to categorical secular versions—which is the same as the objection to the categorical religious versions—is that they do not permit discriminating judgments and are therefore not very helpful in practice. On the other hand, the argument is a reminder to consider the ethical aspects of the biotechnological project very carefully.

B. Violation of the Natural Order

Another intrinsic argument is that genetic modification is "unnatural" or a matter of "violating the natural order." The discussion of this argument is often superficial and unclear on the part of both advocates and critics. I will go a step further and identify different types of unnaturalness. The question is exactly which aspects of the natural order are violated. Another issue is what it means to violate.

The arguments are only to some extent applicable to the generation of genetically modified animals for research. They are commonly used against genetic modification in general.

Midgley, in trying to understand the argument from unnaturalness, writes:

> To say that this change is *unnatural* is not just to say that it is unfamiliar. It is unnatural in the quite plain sense that it calls on us to alter radically our whole conception of nature (Midgley, 2000, p. 12).

Midgley's statement suggests that the argument concerns our view of nature. It is vital to clarify which genetic modification methods are supposed to be unnatural and violate the natural order and which aspects of the natural order are violated by those methods.

Modification through breeding might be viewed as unnatural in the sense that it does not occur without human intervention. It is intentional and directed to particular human ends. Extreme critics may be against all breeding, but most critics probably accept some types. "Unnatural" in this sense consequently does not automatically mean "ethically wrong."

Modification through gene technology is more radical. It involves individual genes instead of the whole genome, as in breeding. The most radical genetic modification method is insertion of genes from species far away. In plant biotechnology, for example, an anti-freeze gene from flounder—a fish living in cold waters—has been inserted into potatoes, in order to make them tolerant to coldness (Jaffé and Rojas, 1994). Such things do not happen in nature, and opponents might use it as a key example of unnaturalness.

Many different versions of the argument of violation of the natural order have been proposed. One focuses on violation of the species barrier. This argument is relevant only for one special form of genetic modification, namely insertion of genes from other species. It is not a matter of "mixing of genomes"; only one gene or a few genes from the foreign species are transferred. The term "transgenic animal" is sometimes used for an animal that contains a gene from another species in all its cells (see Chapter One).

Historically, it is interesting to note that mixed monsters represent threatening disorder. A key example is the chimera of ancient Greek mythology. Chimera was a creature consisting of parts from different animals. As we have seen, "chimera" in modern genetics is the designation for animals at one particular step in the process of producing genetically modified animals with the embryonic stem cell method. A chimera has two different types of cells that are genetically distinct and which originated in different zygotes (fertilized eggs). Some cells contain the new inserted gene, while others do not. Chimeras should not be confused with mosaics, which are also animals with genetically different cell types, but which originate from a single zygote. As we have seen, mosaic animals also constitute one particular step in the process of producing genetically modified animals, although not with the embryonic stem cell method but with the method of pronuclear microinjection.

An objection is that no fixed species barriers exist. According to the theory of evolution, species are not timeless essences in an Aristotelian sense. Some scientists question the whole idea of firm divisions among species. Others stress that there nevertheless is some stability over time (for an overview, see Mishler and Brandon, 1998). In another objection, the very metaphor of trespassing species barriers is challenged. Many forms of genetic modification would only be like adding or deleting furniture in a room. Not only can characteristics be moved about among species, no reason exists in principle why not all characteristics could be so moved.

We also find a weaker interpretation of the species barrier argument. It is non-categorical and implies that crossing species barriers can sometimes be too radical or include too many characteristics. This weaker version appears more reasonable.

Let me also mention three arguments—discussed by Richard Sherlock—that are only to a limited extent relevant to genetic modification of laboratory animals but perhaps more relevant to genetic modification of farm animals. They all focus on different aspects of evolution and maintain that genetic modification would constitute a violation of the natural order that is the result of evolution.

According to the first argument, genetic modification is a violation of the direction of evolution. Transgenics is giving evolution a bad direction. By using a metaphor, it is simply driving the wrong way (*cf.* Sherlock, 2002). An objection is that new species always come into being as transformations of old species by mutations. Evolution has no predetermined direction.

The next argument focuses on the speed of evolution and points out that transgenics is a dangerous speeding up of evolution. The metaphor is one of driving too fast (*cf.* Sherlock, 2002). To this argument it might be objected that the speed varies in evolution too and that speeding is not always dangerous.

Finally, we have the argument that transgenics is a dangerous human steering of evolution, putting natural selection out of rule. Again it is a metaphor of driving, but this time the driver is the wrong one (*cf.* Sherlock, 2002). The most obvious objection to this is that also in conventional breeding, human beings are steering and this is not necessarily dangerous.

All these arguments presuppose that nature is normative. In fact, the notion of "the natural order" is itself a metaphor of legal or social norms applied to nature. Is nature normative? Basically, two different answers have been given. According to one view, nature is value-neutral, according to another it is value-laden. Within the latter category, we find the view that nature is laden with moral values and the view that it is laden with non-moral values. In a previous chapter, I discussed the problem of deriving an "ought" from an "is." I argued that facts about our evolved human nature can be highly relevant for ethics. However, we cannot in a simple way "read-off" any normative conclusions from statements of natural fact. We should be particularly suspicious against all attempts to do so from non-human nature. To take human nature seriously in ethics is one thing, to draw normative conclusions from non-human nature quite another. This was precisely the view of Hume (see Chapter Three).

We must be able to discriminate between different uses of the argument from violation of natural order. Some uses of the argument might be acceptable, while others are not. As Holmes Rolston III points out:

> Critics sometimes object to genetic manipulation because it is "unnatural." "Unnatural" is a dangerous normative term. Most of our cultural activities, such as attending ethics conferences, are unnatural in the sense that they are not found in wild spontaneous nature. Diseases are natural; we seek to heal diseases. Health too may be natural, but medically manipulated pharmaceuticals are cultural artifacts. Whether we can object to an activity as being unnatural is case specific. Sometimes yes, sometimes no, the determining norms may come from culture not nature (Rolston, 2002, p. 10).

Rolston's key idea is that whether we can object to an activity as unnatural is case-specific. I agree. Moreover, as with all intrinsic ethical arguments, categorical and non-categorical versions have been proposed, and the non-categorical seem more useful in practice.

C. Violation of Animal *Telos*

An argument that is quite similar to the violation of natural order argument is that genetic modification of animals is a violation of animal *telos*. This is a metaphor of direction. Like an arrow that has been shot toward a particular goal, animals are by their genetic constitution given a particular direction toward which they are to develop. Hindering this development of their potential, would be like stopping the arrow from reaching its goal.

An example is the argument of Michael Fox, who states:

> Transgenic manipulation is wrong because it violates the genetic integrity or *telos* of organisms or species (Fox, 1990).

Note here the identification of *telos* with genetic integrity. This identification is not self-evident. Both "*telos*" and "genetic integrity" are terms with many possible meanings. Let us focus here on *telos* and leave genetic integrity until the next section.

The key idea of the *telos* argument is that genetic modification, or at least insertion of genes from other species, constitutes a violation of the innate nature of animals. In the quotation, a distinction is indicated between the *telos* of organisms and the *telos* of species. This means that two different versions of the argument exist, one stressing that the individual *telos* is violated, the other that the species-typical *telos* is violated.

A first possible objection is that this argument (in both versions) presupposes an outdated and unscientific conception, namely a conception of a fixed *telos* in an Aristotelian sense. According to Darwinism, no fixed *telos* exists. A possible reply would be that the anti-essentialism of Darwinism is sometimes overstated and that animals have a sufficiently stable species-typical nature to justify the talk about *telos* even within a Darwinian framework.

Henk Verhoog has put forward another objection. He says:

> We misuse the word *telos* when we say that human beings can "change" the *telos* of an animal or create a new *telos* (Verhoog, 1992).

Verhoog appears to mean that a domesticated animal still has its natural *telos*, even if it has been modified by breeding or genetic modification. Its natural *telos* is that part of its nature that is due to its natural endowment.

From an analytic point of view, this means that one more distinction needs to be made besides the one between individual *telos* and species-typical *telos*, namely a distinction between actual *telos* and natural *telos*. The actual *telos* is a *telos* that may have been genetically modified, while the natural *telos* is an unmodified *telos*.

A third objection is that a single new gene does not affect *telos*, for example, "mouseness." A mouse is still a mouse even if it has received a gene from a foreign species.

Finally, we find the objection that genetic modification is not in itself a violation of animals' natures. Whether it is a violation depends on the result of the genetic modification. We saw in the presentation of different animal welfare concerns that Rollin argues for a natural living approach. Animal welfare is a matter of realizing animal *telos* in the sense that all its abilities, including its full repertoire of natural behavior, are expressed. I also indicated that Rollin maintains that adapting *telos* to environment by genetic modification can be ethically acceptable. This is how he argues:

> Let us suppose that we have identified the gene or genes that code for the drive to nest. In addition, suppose we can ablate that gene or substitute a gene (probably *per impossibile*) that creates a new kind of chicken, one that achieves satisfaction by laying an egg in a cage. Would that be wrong in terms of the ethic I have described? If we identify an animal's telos as being genetically based and environmentally expressed,

we have now changed the chicken's telos so that the animal that is forced by us to live in a battery cage is satisfying more of its nature than is the animal that still has the gene coding for nesting. Have we done something morally wrong? I would argue that we have not. Recall that a key feature, perhaps *the* key feature of the new ethics of animals I have described, is concern for preventing animal suffering and augmenting animal happiness, which I have argued involves satisfaction of telos (Rollin, 1995, p. 172).

This means that feeling is supreme in Rollin's animal welfare concept and that animal traits can be genetically modified as long as the animals do not suffer.

Rollin's discussion illustrates that the argument from violation of animal *telos* can be articulated in categorical as well as non-categorical terms. This follows the pattern we have found also with regard to the other arguments expressing intrinsic ethical concerns.

D. Violation of Animal Integrity

Another argument is that genetic modification constitutes a violation of animal integrity. We have already met the argument in the quotation from Fox above: "Transgenic manipulation is wrong because it violates the genetic integrity or *telos* of organisms or species" (Fox, 1990). As we saw, Fox identifies *telos* with genetic integrity. He talks about the genetic integrity both of individual animals and of species. It is crucial to distinguish this genetic or genotypic integrity from phenotypic integrity. While the former concerns the genome—whether species-specific or individual—the latter concerns the individual animal as a whole.

What are the main objections to the argument from genetic integrity? One objection is that what matters ethically is not genetic or genotypic integrity but phenotypic integrity. In Chapter Three, I presented the clarifying analysis by Rutgers and Heeger of animal integrity in the phenotypic sense. They argue that in a state of integrity the following three elements must be present: the wholeness and completeness of the individual animal, the species-specific balance of the creature, and the animal's capacity to maintain itself independently in an environment suitable for the species (Rutgers and Heeger, 1999). With this analysis in mind, the objection is that genetic modification does not necessarily constitute a violation of integrity in any of these senses. Only if the genetic modification violates any of the three elements does it constitute a violation. Surgery or pharmacological treatment may sometimes violate phenotypic integrity even more than genetic modification. Replacing a disease gene with a properly functioning gene may instead be considered a strengthening of phenotypic integrity.

Another objection is that even if we accept the notion of genetic or genotypic integrity, the type of genetic modification is crucial for whether or not the modification is a violation. A genetic modification may strengthen phenotypic integrity but also genetic or genotypic integrity by replacing a disease gene with a properly functioning gene. Only genetic modifications that seriously reduce the function of genes should be considered violations of genetic integrity. This view is an example of a non-categorical view of genetic integrity, in contrast to the view of Fox, which is categorical.

Regardless of whether we talk about genotypic or phenotypic integrity, the distinction between categorical and non-categorical views is extremely important (see Chapter Three). In categorical arguments, integrity is a matter of all-or-nothing, while in non-categorical versions integrity is a matter of more-or-less. Examples of categorical views are Fox's criticism of all genetic modification and Regan's criticism of all animal experimentation including genetic modification. An example of a non-categorical view is Rolston's view that most genetic modifications—but not all—are violations of integrity (Rolston, 2002). Another example is the less dismissive view defended by Donald Bruce. He considers genetic modification unacceptable if the violation of integrity is very substantial, but accepts a balancing of pros and cons regarding the rest (Bruce, 2002).

In conclusion, the integrity argument has a point, but integrity should not be viewed as a matter of all-or-nothing but of more-or-less. Sometimes genetic modification is so radical that it violates the integrity of animals, sometimes not. Moreover, what truly matters is phenotypic integrity, not genotypic. It is not the genetic modification as such that counts, but the result on the phenotypic level.

E. Special Arguments regarding Animal Cloning

The above arguments may all be used against animal cloning, in categorical and non-categorical versions. However, they will have a special character, because cloning by nuclear transfer involves asexual reproduction in animal species characterized by sexual reproduction. This is especially obvious with regard to the argument from violation of the natural order. Asexual reproduction in a species with sexual reproduction is regarded as unnatural. As such it could also be viewed as a violation of animal *telos*, because the animal is not given the opportunity for sexual reproduction, which would be a part of its genetic constitution. Asexual reproduction may also be considered "playing God," because it would imply choosing a mode of reproduction for a species, something that only God can choose.

What about the argument from violation of animal integrity? In categorical versions, genotypic integrity may be viewed as violated, because sexual reproduction has a firm genetic basis. This holds true also of categorical versions of phenotypic integrity, since it is part of the species-typical phenotype to reproduce sexually. From a non-categorical viewpoint, according to which genotypic or phenotypic integrity is not a matter of all-or-nothing but of more-or-less, animal cloning could perhaps be accepted, provided that it is carried out for good reasons and provided that it concerns asexual reproduction only or that it is combined with another acceptable genetic modification.

The possible objections to these arguments are the same in the case of animal cloning as in the cases of other types of genetic modification, but "unnaturalness" requires a special comment. It is obvious that animal cloning by nuclear transfer is unnatural in an everyday sense that anybody can recognize. It is a matter of asexual reproduction in species whose species-typical mode of reproduction is sexual. However, in the argument from violation of the natural order, naturalness is considered to have a normative force that requires us not to carry out this type of genetic modification.

F. Concluding Comment regarding Intrinsic Ethical Concerns

The only intrinsic ethical argument that carries real weight is a non-categorical version of the argument from violation of animal integrity focusing on phenotypic integrity. Some genetic modifications may constitute a violation of integrity in this sense. This means that the burden of proof lies on those who want to carry out the modification. They have to show that the modification is sufficiently beneficial to outweigh the violation. Violation of phenotypic integrity is a cost that should be included in the balancing of human benefit and animal harm, although it is a very difficult cost to quantify.

4. Animal Welfare Concerns

In the previous chapter, I analyzed the concept of animal welfare and its implications for animal experimentation in general. Now, I will discuss the animal welfare implications for genetically modified animals in research.

We have identified three different animal welfare concerns, namely functioning, feeling, and natural living. Which one of these we primarily focus on may be of vital importance for the assessment of the welfare of genetically modified animals. Below, I will show how different animal welfare concerns may differ with regard to the welfare of genetically modified animals. I will also comment on the implications of experiments involving genetically modified animals for the 3Rs.

The overall aim is to discuss four different aspects of the welfare of genetically modified animals. The first is the welfare of animals used in the process of production. The second concerns the welfare of the resulting genetically modified animals, which are to be used in further studies. Third, some animal welfare aspects are related to the preservation of the generated genetically modified animals. Finally, some welfare issues are raised by the experimental use of genetically modified animals.

I have suggested that the three animal welfare concerns be attributed different weight in different contexts. With regard to laboratory animals, feeling in terms of absence of pain is the most important concern, although stimulation of positive feelings by an enriched environment is also very important. Functioning well in terms of health is important, but if the animals are to be used as disease models, it would have low priority. Since the animals have to be kept in cages for reasons of scientific control, natural living would have low priority. Regarding farm animals, functioning and feeling are the most important animal welfare concerns, while natural living has lower priority (although it may be of some instrumental importance). Since the animals are used for production, they should be in good health and not suffer. They have to be kept in confined areas and thereby be protected from predators.

These considerations hold true also of genetically modified laboratory animals and genetically modified farm animals. Even in these cases, the different animal welfare concerns may be attributed different weight in a similar manner. In the case of knock-out studies of gene function, the concern of functioning well will have as low a priority as in studies of disease models.

A. The Production Process

Let us start with the animal welfare aspects of the process of production of genetically modified animals (Moore and Mepham, 1995; CCAC, 1997; van der Meer, 2001; BVAAWF *et al.*, 2003). I focus on the two main methods: pronuclear microinjection and the embryonic stem cell method. Both have been described in the section on scientific concerns.

Three of the steps seem to involve only minor negative effects on animal welfare. Hormone stimulation of donor females leading to superovulation might be stressful, but should not be overstated. The killing of donor females—in mice commonly by dislocation of the neck—is a sensitive measure, but if it is carried out by competent and experienced personnel, the procedure is painless and death immediate. Vasectomy of males is another sensitive measure. It is carried out with anesthesia and is not followed by any serious post-surgical pain. The degree of harm of these steps should reasonably be classified as mild with regard to pain and distress. Killing, however, even if it is painless, is a violation of animal integrity, because it deprives the animal of the opportunities of life. This violation of animal integrity should be included as an additional cost in ethical balancing.

Surgical implantation of embryos into pseudo-pregnant females is a more serious intervention from an animal welfare perspective. It is carried out with anesthesia, but should nevertheless be classified as moderate with regard to pain and distress. It is a matter of major surgery in the stomach cavity.

A special animal welfare problem is raised by embryo implantation and vasectomy in mice. Mice are prey species and try to conceal or suppress signs of pain or stress in order not to appear to be an easy prey and attract the attention of predators. This means that mice after surgical intervention may not exhibit any clear signs of suffering. With this in mind, all mice undergoing surgery or other potentially painful procedures should receive analgesia (BVAAWF *et al.*, 2003, p. S1:20).

Other welfare issues are raised by genotyping. In order to distinguish genetically modified mice from non-modified ones, it may be necessary to analyze DNA extracted from a tissue sample or blood. Often this is done by a tail biopsy or by ear notching, but there may also be non-invasive options such as the use of saliva or fecal samples. Southern blot hybridizations require more DNA than polymerase chain reaction (PCR). If tail biopsies are unavoidable, anesthesia and analgesia are to be used and no more than 5 mm of the tail should be taken. From an animal welfare perspective, ear notching is preferable because less tissue is removed and the pinna is completely cartilaginous.

For many research protocols, it is necessary to identify individual genetically modified animals. Even here non-invasive methods are preferable from an animal welfare standpoint. Often, however, invasive methods are necessary. Examples are ear notching, ear tags, microchips, and tattoos. In these cases, anesthesia and analgesia should be used in order to minimize pain and distress.

In this discussion I have focused on feeling only. This is in line with what I have previously stated, namely that in the case of animal experimentation feeling is the most important animal welfare concern.

Let me make a few remarks regarding the production process in animal cloning. The efficacy of animal cloning by somatic cell nuclear transfer is low. The first cloned mammal—the well-known sheep Dolly—was the result of 277 trials (Wilmut *et al.*, 1997). This tendency is still the same ten years later, but the efficacy may vary among different species. Snuppy, the first cloned dog, was the only survivor—of two live births—out of 1,095 trials (Lee *et al.* 2005). A conclusion is that an overwhelming majority dies at the embryonic or fetal stages. I will return to the problems with animal cloning below.

B. The End Result: Genetically Modified Animals

The welfare of the resulting genetically modified animals is of key importance (*cf.* Moore and Mepham, 1995; CCAC, 1997; van Zutphen and van der Meer, 1997; van der Meer, 2001). It is vital to recognize that the welfare implications might differ considerably for genetically modified animals.

Some genetically modified animals might exhibit improved welfare. Farm animals, for example, may be genetically engineered to have improved disease control. This might reduce the use of vaccines, pharmaceuticals, quarantine, and selective breeding, which may sometimes be distressing. Farm animals may also obtain improved disease resistance through genetic modification. In practice, this might be difficult to achieve, because in many diseases several different genes may be involved.

Another possibility could be to produce animals with reduced sentience or cognitive capacity. In this way, low welfare in terms of suffering may be avoided. Laboratory mice could be generated that are unable to feel pain. Battery chickens could be produced that feel no urge to peck or stretch their wings. Whether this should be viewed as improved welfare depends on which animal welfare concern we stress. If the focus is on feeling, then it could be considered an improved welfare. If the focus is on natural living, it may not. If the animal's sentience is reduced to the extent that it may be considered to be a mere instrument or artifact, then it cannot manifest its *telos* or nature. Many people view precisely the ability to feel pain as a characteristic that confers moral relevance to animals and would not accept a reduction of this capacity.

The possibility also exists that animal welfare is unaffected. Most genetically modified bioreactors, especially those producing therapeutic proteins in their milk, would not have their welfare affected at all. The reason is that milk and its protein contents are isolated from the other tissues of the animal. Moreover, scientists have developed genetically modified disease models with no clinical symptoms. This represents an ethical advantage compared to non-modified disease models.

A substantial proportion of genetically modified animals appears to have reduced welfare. Estimating the number is difficult, but some researchers believe that the figure is less than 10% (BVAAWF *et al.*, 2003, p. S1:35). However, in our Swedish analysis of applications submitted to ethics committees on animal experimentation, we found that obvious or minor clinical symptoms due to genetic modification were expected in more than a third of the applications (Nordgren and Röcklinsberg, 2005). We may compare this finding with the result of a study of reports to the Danish Animal Experiments Inspectorate. In this study, it was found that 36% of the genetically modified strains were reported as experiencing discomfort (Thon *et al.*, 2002). Clinical symptoms and experienced discomfort are different things, so are estimations before experiments and reports afterwards. Moreover, a focus on number of applications differs from a focus on number of animal strains. Consequently a direct comparison is not possible. However, both studies point in the same direction, namely that a substantial portion of genetically modified animals may have a quite bad welfare (Nordgren and Röcklinsberg, 2005).

In some cases, reduced welfare is compensated for in different ways. Genetically modified farm animals may have increased productivity, but this may be accompanied by more production diseases. For example, cattle with greater milk yields may be susceptible to mastisis, although this condition may be treated by medication. Genetically modified sheep with increased wool growth may have their thermoregulatory capacities affected in a negative way, but it might be possible to compensate for this by environmental control.

In other cases, reduced welfare of genetically modified animals is an intrinsic part of the experiment. One example is gene knock-outs with known or unknown effects. Such mice are used in studies of gene function. Another example is the use of genetically modified mice as disease models. These models may exhibit clinical symptoms and experience pain. If this does not interfere with the researchers' scientific objectives, they might be given pain relief and have their environment enriched. Moreover, if they are used for testing new treatments and these are successful, then suffering may be relieved and organs may function properly again. A third example is genetically modified animals used in toxicity testing. Here the objective is precisely to see at what point a particular substance gives toxic effects. All these examples of welfare are primarily conceived of in terms of functioning. The aim is to have measurable negative effects on functioning. This means, on the other hand, that functioning well has low priority regarding these animals. It is quite possible, however, to give welfare in terms of feeling well high ethical priority by providing pain relief.

It is crucial to be aware of different animal welfare concerns. For example, a genetically modified animal working as a model for cancer at an early stage has a health problem in terms of bad functioning but may not yet feel pain (Buehr et al., 2003). If an animal has been genetically modified, it may exhibit unnatural behavior but may still function well. In many cases, the different concerns overlap. For example, a genetically modified animal at a later stage of cancer may have low welfare on all accounts.

Sometimes genetic modification methods might be better from an animal welfare perspective than non-genetic ones. For example, genetically modified diabetes models may be preferable to pharmacologically or surgically induced models in this regard. Let me also mention an advantage of conditional methods. Tissue-specific and temporally specific modification provides a mechanism for minimizing negative effects on animal welfare (BVAAWF et al., 2003, p. 6).

In sum, the welfare of genetically modified animals may vary. Genetically modified farm animals might exhibit improved welfare. Genetically modified animals to be used as bioreactors or providers of organs for xenotransplantation will commonly show no change in welfare but will be very well taken care of. Genetically modified disease models may exhibit the whole spectrum from no clinical symptoms whatsoever to severe clinical symptoms and the whole spectrum from no suffering to severe suffering. The same may hold true for genetically modified animals to be used as study objects for obtaining basic biological knowledge. Genetically modified animals to be used for toxicity testing can often be expected to exhibit obvious clinical symptoms and suffering.

So far I have talked about intentional reduction of welfare. Another possibility is unintentional reduction. Here the method of production is relevant. Such welfare effects occur primarily when pronuclear microinjection is used, which is characterized by random integration, although these effects can be reduced by means of insulator or intronic sequences. In the embryonic stem cell method, random integration is not an issue, because this method is based on homologous recombination carried out *in vitro*; in principle, only embryonic stem cells with the desired genetic modification are injected into the blastocyst, while those with insertional mutations are not. Because of this fact, unintended effects of using this type of method are rare in living animals (Buehr et al. 2003). It appears, therefore, that the embryonic stem cell method has an ethical advantage with regard to unintentional animal welfare effects compared to pronuclear microinjection (Nordgren and Röcklinsberg, 2005).

Even if the embryonic stem cell method is used, unintentional effects may occur. Even if a gene is correctly inserted, the animal carrying it may exhibit an unexpected phenotype (Buehr *et al.*, 2003). This may partly be due to epigenetic factors (see below). Colin J. Moore and T. Ben Mepham have pointed out that

> epigenesis therefore limits the validity of attempts to genetically engineer animals. Few characteristics can be modified or introduced, reliably and predictably, by manipulation of a single gene. This has implications, particularly for the effectiveness of transgenic disease models and consequently on the number of animals generated in the quest for their successful production (Moore and Mepham, 1995, p. 391).

Curt D. Sigmund has also argued that animals containing the same genetic modification may exhibit profoundly different phenotypes because of epigenetic effects. In the absence of standardized inbred mouse strains, no optimal set of experimental and control conditions exists that normalizes epigenetic effects. With this in mind, Sigmund suggests that it becomes the responsibility of the investigator to use common sense and design the best possible control experiments that fit the individual situation, to assess whether the phenotype observed in their model is due specifically to the targeted modification or is affected by other loci, and to inform the scientific community if phenotypic alterations become evident (Sigmund, 2000).

However, unpredictable welfare effects appear in all animal experimentation, and genetic modification methods may often make the outcome more predictable, for example when compared to selective breeding.

Finally, difficulties may arise in detecting clinical symptoms, and more research is needed. A practical and quite promising method is welfare scoring (Crawley, 2000; van der Meer *et al.*, 2001; Jegstrup *et al.*, 2003).

Let me also comment on the animal welfare aspects of cloning. Even here epigenetic effects are important (*cf.* Nordgren, 2006).

I mentioned above the very low success rate in animal cloning. Only very few cloned animals are born given the high number of trials, and many of the clones that are born exhibit abnormalities such as respiratory distress, circulatory problems, immune dysfunction, kidney failure, and brain failure. Many cloned animals are overgrown. This condition is often called the "large offspring syndrome" (Rideout *et al.*, 2001, p. 1095). In cloned mice, obesity is common (Fulka *et al.*, 2004).

The problems in animal cloning are increasingly believed to be due to errors in epigenetic reprogramming (National Academy of Sciences, 2002; Humpherys *et al.*, 2002; Alberio and Campbell, 2003; Jaenisch and Bird, 2003). Epigenetics concerns chemical changes that "switch" genes on or off. This epigenetic regulatory information is not expressed in DNA sequences but transmitted to the next generation of cells "in addition to" (*epi*) the genetic information encoded in the DNA. All cells have the same set of genes, although which genes are active and which are not vary from one type of cell to another. In animal cloning, the genetic "switch" must be reprogrammed from the adult state to the gamete or embryo state. William M. Rideout *et al.* maintained that

> cloning of mammals by nuclear transfer (NT) results in gestational and neonatal failure with at most a few percent of manipulated embryos resulting in live births. Many of those that survive to term succumb to a variety of abnormalities that are likely due to inappropriate epigenetic reprogramming (Rideout *et al.*, 2001, p. 1093).

An indication that the abnormalities are epigenetic instead of genetic is that neither obesity in cloned mice nor the "large offspring syndrome" in cattle are passed on to offspring. They are observed only in the founder generation (Fulka *et al.*, 2004).

In order to explain this fact, Fulka *et al.* suggested that epigenetic reprogramming occurs in two steps. The first step occurs during the first divisions after nuclear transfer. In about 1–5% of the cases, this reprogramming—perhaps with some errors—results in viable offspring. In the remaining cases, the reprogramming is incomplete with the result that embryos die or offspring are not viable. During the second step, imprinted and non-imprinted genes are reprogrammed, and errors that were not repaired during the first step are corrected. This step occurs only in germline cells. So, while cloned animals may contain somatic cells with abnormalities, their spermatozoa or oocytes have no errors (Fulka *et al.*, 2004).

If the explanation suggested by Fulka *et al.* is correct, this would have important practical implications. First, cells obtained by human therapeutic cloning may have abnormal gene expression caused by epigenetic errors. Second, the problems of incomplete epigenetic reprogramming would constitute an important argument against human reproductive cloning. Third, the fact that the offspring of cloned animals will be normal would be vital for the use of cloned animals in xenotransplantation and the production of pharmaceutical proteins in their milk (Fulka *et al.*, 2004).

It is still possible to object that perhaps Fulka *et al.* are not entirely correct. We simply do not know for certain. It might still be possible to improve methods to reduce the risks of epigenetic errors. For example, time might be important. If so, it may be possible to delay cell division in clones, giving time for proper reprogramming to occur. In addition, exogenous factors—for example cell culture conditions—may determine the outcome, and these may be modified (Simpson, 2003).

C. Preservation of Genetically Modified Animals

When the production of genetically modified animals has been carried out, the next problem is to preserve. If no reduced welfare effects have been observed in homozygotes, these are preferably maintained. This prevents production of surplus mice with an undesired genotype. If homozygotes experience reduced welfare, heterozygotes should be maintained instead. These have the deleterious allele only in one copy and may not suffer at all. An even better option could be cryopreservation of gametes, ovarian tissue, or early stage embryos. This reduces the number of animals in animal houses.

D. Experimental Use of Genetically Modified Animals

In the experimental use of already produced genetically modified animals, the problem of reduced welfare due to genetic modification is quite different than in the production of such animals. The researchers may be informed by colleagues from whom they obtain the animals or by the company from which the animals are bought.

We have already discussed welfare aspects of the production of genetically modified animals with different intended uses. What happens to the animals once they are produced differs.

Some genetically modified animals are treated with extraordinary care because of their high economic value. This holds true for bioreactors, providers of organs for xenotransplantation, and farm animals with increased productivity.

Disease models, on the other hand, often have low welfare for reasons intrinsic to researchers' scientific objectives. Some may be used without exhibiting any clinical symptoms. The welfare of disease models may be changed because of their specific experimental use. If the experiments aim at studying disease processes and causes, their welfare may be reduced even more over time. If the aim is to develop or test new treatments, their welfare may be improved over time should the treatment prove effective.

The welfare of genetically modified animals used in toxicity testing may initially be unaffected or reduced depending on the genetic modification. However, due to the effects of toxication, it may be reduced or, if initially reduced, even more reduced.

Genetically modified animals used as study objects for obtaining basic biological knowledge may exhibit the whole spectrum of welfare depending upon the type of genetic modification. As already stressed, mice used in knock-out or overexpression experiments may suffer. Genetically modified animals may also be used to study other biological processes, and their welfare may differ, depending on the type of study.

In general, the welfare aspects of experimental use of genetically modified animals are the same as those in "ordinary" animal experimentation: alternatives, design, species, numbers, and so on (see Chapter Five).

E. The 3Rs and Genetic Modification

What are the implications of using genetic modification methods for the 3Rs? And what are the implications of the 3Rs for the use of genetic modification methods? These are two questions that remain to be answered with regard to the welfare concerns of genetically modified animals.

In the section on scientific concerns, I discussed the necessity and suitability of genetically modified animals in research. To the extent that genetic modification methods are necessary or better than non-genetic ones, it is hardly possible to replace them with non-animal alternatives (Moore and Mepham, 1995). However, to some extent it might be possible to replace farm animal bioreactors with microorganisms.

Contrary to the principle of reduction, more animals will probably be used in experimentation due to genetic modification (Stokstad, 1999). It is obvious that large numbers of mice are used in the process of producing those animals that are to be used in experiments. Moreover, many of the embryos do not survive, and of those born relatively few are genetically modified (1–30%, on average 15%). Thus, there will probably not be a reduction in absolute numbers, although there might be a reduction in relative numbers. Fewer animals will be used in order to answer a specific scientific question (Buehr *et al.*, 2003).

A possibility of refinement exists by producing genetically modified animals with reduced sentience or cognitive capacity, for example, laboratory mice that are unable to feel pain or battery chickens that feel no urge to stretch their wings or peck. Whether this should be viewed as refinement depends upon which animal welfare concept you use. If the focus is on feeling, this might be considered an improvement of animal welfare. If the focus is on natural living, it would not, because reduced capacities would be a matter of violating the nature or *telos* of the animals. Moreover, the possibility of unpredictable effects that could cause animals to suffer is contrary to refinement (Moore and Mepham, 1995).

The considerations so far have concerned genetic modification methods in comparison with non-genetic ones. Let me give a few examples of how scientists using genetic modification methods can meet the 3Rs.

Some uses of genetic modification methods may create possibilities for the replacement of animal experiments. For instance, it might be possible to use genetically modified animal tissue expressing a particular gene product. This will thereby give the opportunity to perform further experiments largely *in vitro* (Buehr *et al.*, 2003).

Cryopreservation may lead to reduction of the number of animals used. It will not be necessary to keep animals in animal houses for extended periods of time (Moore and Mepham, 1995).

Finally, some refinements are possible (see above). Adverse effects of random integration may be reduced by inclusion of insulator or intronic sequences in the transgene. Conditional methods may minimize bad animal welfare effects due to genetic modification, since they make these modifications tissue-specific or temporally specific. Many refinements in surgical techniques and genotyping are possible. Several refinements like these are suggested in the article by the BVAAWF *et al.* (2003, pp. S1:44–46).

5. Ethical Trade-Off: Four Cases

Let us consider the ethical trade-offs in a few particular cases. In this way the discussion will be made more concrete. I will present four cases of genetic modification of animals that are fairly typical and also illustrate different types of ethical problems raised by this kind of research. Some of these cases are particularly difficult, since they involve moderate or severe animal pain. Ethical reasoning concerning these four cases will illustrate how the ethical matrix model presented in the previous chapter can be applied. The cases are real cases from recent articles in scientific journals. However, in these articles the animal experiments are reported after they have been carried out. Scientists and ethics committees on animal experimentation, on the other hand, are to ethically deliberate on such experiments before they are carried out. With this in mind, I will discuss the ethical trade-off from both perspectives. This gives us an opportunity to investigate possible differences in assessment depending on the temporal perspective.

A. Case 1: Pronuclear Microinjection

The first case is an animal experiment in which pronuclear microinjection is used to transfer a gene for human erythropoietin (hEPO)—a human growth hormone—into mice (Kim, Kim, Shin *et al.*, 2007).

Background. EPO is the primary regulator of erythropoiesis, that is, the formation of mature red blood cells. It binds to a receptor on erythroid progenitor cells in the bone marrow and stimulates cell proliferation, promotes cell differentiation, and prevents cell death. The expression of the EPO gene depends on developmental stage and tissue type. During fetal development the main source of EPO is the liver. From late in gestation onwards the kidney is the major producer. Under anemic stress in adult life the liver may also contribute. Patients with chronic renal failure develop anemia due to inadequate production of EPO by the kidneys. Genetically engineered EPO may be used to eliminate the need for blood transfusions. The patients get regular injections 2 to 4 times a week. hEPO is produced in Chinese hamster ovaries. Several attempts have been made to express EPO in transgenic mice. Many attempts have also been made to produce hEPO in the milk of genetically modified mice, although in most cases the expression of the protein was very low and the animals showed unexpected clinical symptoms (Kim, Kim, Shin *et al.*, 2007).

General design of the experiment. In the present project, transgenic mice were generated that express hEPO under the control of beta-casein regulatory sequences. EPO expression vectors were constructed in the laboratory. The DNA construct was microinjected into fertilized eggs. Five out of 21 mice were used as founders. They were identified by PCR analysis of DNA obtained by tail biopsy. Blood was drawn by the eye-bleeding method and the concentration of hEPO was measured. Histological analyses were also carried out (Kim, Kim, Shin *et al.*, 2007).

Result. High expression of hEPO was achieved in lungs and liver and lower expression in kidney and spleen. The founders exhibited serious disease symptoms such as lung failure, liver failure, tumors, and erythrocytosis. They also had such a short life span due to these symptoms that the scientists failed to make them pregnant. It was concluded that it was the secreted hEPO that caused these deleterious effects (Kim, Kim, Shin *et al.*, 2007).

Human benefit. A positive result is that it is possible to produce large amounts of hEPO in genetically modified mice.

Animal harm. Although it is possible to produce large amounts of hEPO in mice, this will have serious negative effects on animal welfare. Several preparatory steps in the generation process are not mentioned in the article, such as superovulation, mating, vasectomy, and so on. We have seen above that all these steps have implications for animal welfare. What is most important in this experiment is that the genetically modified mice develop lung failure, erythrocytosis, and other disease symptoms and, as an effect of this, have very short lives. On any conception of animal welfare—feeling, function, or natural living—the welfare impact on the animals is severe. In addition, the cost in terms of violation of animal integrity is very high.

Ethical trade-off. This animal experiment shows that high expression of hEPO in genetically modified mice can be achieved, but also that this is combined with serious clinical effects on the mice. This knowledge suggests that mice should not be used for production of hEPO. Other options need to be investigated. These are conclusions that can be drawn after the experiment has been carried out. Should it have been approved in advance? Indications in previous experiments of low expression and severe animal harm might have suggested that the experiment should not have been approved. However, knowledge about whether it is possible to achieve high expression and more well-founded knowledge about whether this would have severe effects on animal welfare would be crucial. This suggests that the experiment should be approved also in advance.

B. Case 2: Experimental Use of Knock-Out Mice

In this case, knock-out mice are used as models for the pathophysiological mechanisms of pain (Kim, Kim, Back *et al.*, 2007).

Background. Neuropathic pain is a chronic condition characterized by pain responses to non-noxious stimuli, exaggerated pain responses to noxious stimuli, and spontaneous pain. It is caused by injuries in the central or peripheral nervous system. It is often difficult to manage, because it frequently becomes intolerant or refractory to analgesic drugs or surgery. With this in mind, it is important to investigate its underlying mechanisms and possible therapeutic treatments (Kim, Kim, Back *et al.*, 2007).

General design of the experiment. As we saw in Chapter Five, nociception is the activity of nociceptors. It is not a mental state but the first stage in a process that often includes pain. The cyclic AMP second messenger system is involved in nociception, and inhibition of this pathway by blocking the activities of andenylyl cyclase (AC) and protein kinase A prevents chronic pain in animal models. The knowledge of which of the 10 isoforms of AC are involved in nociception is limited. In the present experiment, the potential pronociceptive function of andenylyl cyclase-5 (AC5) is investigated. This is done by using AC5 knock-out mice and comparing their behavior with wild-type AC5 mice as control. Several different pain tests are carried out. The first type of these behavioral tests consists of mechanical pain tests. Here the mechanical sensitivities of hindpaws and tails were measured. Another type is thermal pain tests measuring tail-warm sensitivity, paw-cold sensitivity, and paw-infrared-heat sensitivity. A third category is inflammatory pain tests including a subcutaneous formalin test and a visceral pain test (Kim, Kim, Back et al., 2007).

Results. It is demonstrated that AC5 is necessary for the nociceptive pathways of different types of pain: physiologic pain, that is, acute mechanical and thermal pain, and pathologic pain, that is, inflammatory and neuropathic pain. In particular, the inflammatory and neuropathic pain suppression in the AC5 knock-out mice is intriguing, given that these pains are difficult to manage by conventional therapeutic methods (Kim, Kim, Back et al., 2007).

Human benefit. The results are very important for understanding the underlying mechanisms of chronic pain and therapeutic strategies for managing such pain. The possible human benefit must be expected to be very high.

Animal harm. The wild-type controls exhibit different degrees of pain responses, while the knock-outs show "markedly attenuated" pain responses. This shows that the welfare impact on the animals is diverse. The controls suffer during the experiments, while the knock-out mice benefit from the genetic modification and suffer much less. The integrity of the knock-out mice is strengthened because of the modification instead of violated by it; they function better and with less pain than the controls.

Ethical trade-off. This experiment is likely to be an important step to great medical benefit. Despite the low welfare of the controls, the experiment appears justified from an ethical point of view. This is clearly so when looking back after the experiment has been carried out, but I would probably make a similar assessment in advance; the expected human benefit is so high.

C. Case 3: Conditional Knock-Out

In our third case, conditional knock-out mice are produced and used to understand the roles of a receptor in the adult brain (Nakao et al., 2007).

Background. Metabotropic glutamate receptors are involved in the synaptic transmission and plasticity in the central nervous system. The metabotropic glutamate receptor-subtype 1 is strongly expressed in Purkinje cells of cerebellum. In previous experiments, metabotropic glutamate receptor-subtype 1 knock-out mice show symptoms like ataxic gait and motor discoordination. The development was impaired because of the defect of synapse elimination during the third postnatal week. Other experiments show that this type of knock-out mice were rescued—for example regained motor coordination—by the introduction of a metabotropic glutamate receptor-subtype 1 transgene. The question remains what roles the metabotropic glutamate receptor-subtype 1 plays in the adult brain. In order to investigate this, conditional knock-out mice are generated. In these mice, the receptor is expressed only in the Purkinje cells and its expression can be turned on and off at will (Nakao *et al.*, 2007).

General design of the experiment. The conditional metabotropic glutamate receptor-subtype 1 knock-out mice are generated by using the tetracycline-controlled gene expression system. The expression is controlled by oral administration of a tetracycline analog, doxycycline. Founder mice of two types are generated, one with a transgene expressing a tetracycline-controlled transactivator and one with the tetracycline responsive element. The founders were identified by Southern blot analysis and PCR on tail biopsies. The founder mice are crossed, yielding metabotropic glutamate receptor-subtype 1 conditional knock-out mice with both transgenes. These mice were administered doxycycline in drinking water, and their behavior was investigated by two types of test: a rotating rod task and an ink footprints test (Nakao *et al.*, 2007).

Results. Mice with the metabotropic glutamate receptor-subtype 1 turned off fell off the rod immediately, while those with the receptor turned on learned quickly how to keep themselves on the rotating rod. The footprint test showed similar results. Those with the receptor turned off showed abnormalities in gait and steps and those with it turned on had no problems. In order to exclude the possibility that the doxycycline itself was responsible for the difference, this substance was also administered to wild-type mice with no effect. The results suggest that metabotropic glutamate receptor-subtype 1 is crucial for motor coordination in the adult mice (Nakao *et al.*, 2007).

Human benefit. The animal experiment provides knowledge about the role of the metabotropic glutamate receptor-subtype 1 in the adult mouse brain. This knowledge might be of high human benefit in the long run in understanding and developing treatments for metabolic disorders.

Animal harm. The behavioral tests imply some distress to the mice with the receptor turned off. But this distress is not severe. The welfare impact is moderate and limited in time. The cost of violation of animal integrity appears also only moderate. When the receptor is turned on, they do not differ from phenotypically normal mice.

Ethical trade-off. The conditional knock-out mice survive to adult age making it possible to carry out experiments that would otherwise be impossible. The expected high human benefit outweighs the moderate distress experienced by some of the animals in the behavioral tests. This holds true when looking back and when trying to assess the experiment in advance.

D. Case 4: Animal Cloning

Our final case concerns cloning of dogs by means of somatic cell nuclear transfer (Lee *et al.*, 2005).

Background. Animal cloning was for a long time unsuccessful with regard to dogs. The reason is the difficulty of maturing oocytes *in vitro*. One benefit of acquiring the ability to clone dogs—together with results from the canine-genome project—would be knowledge about the relative contributions of genes and environment to the diversity of physiological and behavioral traits of different breeds (Lee *et al.*, 2005).

General design of the experiment. Oocytes matured *in vivo* and were collected by laparatomy three days after ovulation. During laparatomy the females were anesthetized. The oocytes were enucleated with a micromanipulator. Adult cells were isolated from an ear skin biopsy of a three-year old Afgan Hound and microinjected into each enucleated oocyte. The couplets were fused, and the successfully fused couplets were chemically activated. The embryos that were reconstructed in this way were surgically transferred into the oviduct or the uterus of anesthetized surrogate mothers. Pregnancies were detected by ultrasound. Tissue samples were taken from the tail of the clones, and blood samples were obtained from the donor and the surrogate mother. A DNA analysis was carried out with eight canine specific markers (Lee *et al.*, 2005).

Results. 1,095 reconstructed dog embryos were transferred into 123 surrogate mothers resulting in the birth of two cloned puppies. The first cloned dog was called Snuppy (for Seoul National University Puppy). The other died on day 22 due to pneumonia. The DNA analysis confirmed that they were genetically identical with the donor dog (Lee *et al.*, 2005).

Human benefit. This experiment—together with the canine-genome project—might increase our basic biological knowledge about the interaction of genes and environment in different breeds. It may provide better knowledge of problems related to animal cloning in general. In addition, it may give knowledge of importance to human therapeutic cloning. It may also give knowledge pertinent to human reproductive cloning, although this type of cloning is extremely controversial and by many considered not to carry any human benefit whatsoever, but quite the opposite (Nordgren, 2006).

Animal harm. The success rate was very low, only two dogs were born out of 123 recipients (1.6%). One of these dogs died of pneumonia after three weeks. Low success rate and abnormalities are common experience in animal cloning. These are increasingly believed to be due to incomplete epigenetic reprogramming. The pneumonia is a clear example of low animal welfare. We may interpret also the low success rate as an example of low welfare; the miscarriages indicate abnormalities in embryos and fetuses. However, embryos and fetuses do not count as "animals" in the European Union regulation (see Chapter One). The integrity of the born dogs might be violated only in the sense that they are (almost) identical with the donor—they are not unique—but this does not seem particularly serious.

Ethical trade-off. The low success rate and the pneumonia of one of the dogs followed by its early death are very serious. If we look only at the one successful cloning, the benefit hardly outweighs the cost. From a broader perspective the experiment might nevertheless be ethically acceptable. The basic biological knowledge gained from this experiment may be important, and the application of animal cloning to human therapeutic cloning holds much hope, although it is quite controversial. The application to human reproductive cloning, on the other hand, is extremely controversial. Elsewhere, I have analyzed an "epigenetic argument" against human reproductive cloning that takes the problems in animal cloning seriously (*cf.* Nordgren, 2006). I find no difference in assessment looking at this case in advance compared to afterwards.

6. Conclusion

After this brief discussion of a few cases, we may ask: Is the production and use of genetically modified animals in biomedical research on the whole ethically acceptable or not? From the perspective of the weak human priority view, it is not possible to answer this question in general terms. We can only say that to the extent that such experimentation can be expected to lead to significant human benefit in terms of basic biological knowledge and medical development, and to the extent that it does not inflict too much harm on the animals, it is acceptable. We need to make a trade-off from case to case. Not all production and use of genetically modified animals is acceptable. In some cases, the expected animal harm may outweigh the expected human benefit. These animal experiments should not be carried out.

WORKS CITED

Aiba, Atsu, and Harumi Nakao. (2007) "Conditional mutant mice using tetracycline-controlled gene expression system in the brain," *Neuroscience Research*, 58:2, pp. 113–117.

Alberio, Ramiro, and Keith H. S. Campbell. (2003) "Epigenetics and nuclear transfer," *Lancet*, 361, pp. 1239–1240.

Alcock, John. (2001) *The Triumph of Sociobiology*. Oxford: Oxford University Press.

Aldhous, Peter, Andy Coghlan, and Jon Copley. (1999) "Let the people speak," *New Scientist*, 162, pp. 26–31.

Animals (Scientific Procedures) Act. (1986) The United Kingdom.

Animal Welfare Act. (1985) USA: Department of Agriculture.

Animal Welfare Act ("Djurskyddslagen") 1988:534 (with later revisions and supplements). Sweden.

Animal Welfare Ordinance ("Djurskyddsförordningen") 1988:539 (with later revisions and supplements). Sweden.

Appleby, Michael C., and Barry O. Hughes, eds. (1997) *Animal Welfare*. Wallingford: CAB International.

Appleby, Michael C., and Peter Sandøe. (2002) "Philosophical debate on the nature of well-being: Implications for animal welfare," *Animal Welfare*, 11, pp. 283–294.

Aristotle. (1971) *The Nicomachean Ethics of Aristotle*. Translated by Sir D. Ross. London: Oxford University Press.

Armstrong, Susan J., and Richard G. Botzler, eds. (2008) *The Animal Ethics Reader* (2nd ed.). London: Routledge.

Arnhart, Larry. (1998) *Darwinian Natural Right: The Biological Ethics of Human Nature*. Albany: State University of New York Press.

Augustine. (1877) *The City of God*. Translated by Marcus Dods. Edinburgh: T. & T. Clark.

Aureli, Filippo, and Frans B. M. de Waal, eds. (2000) *Natural Conflict Resolution*. Berkeley: University of California Press.

Barnard, C. J., and J. L. Hurst. (1996) "Welfare by design: The natural selection of welfare criteria," *Animal Welfare*, 5, pp. 405–433.

Bateson, Patrick. (1986) "When to experiment on animals," *New Scientist*, 109, pp. 30–32.

Beauchamp, Tom L., and James F. Childress. (2009) *Principles of Biomedical Ethics* (6th ed.). Oxford: Oxford University Press.

Bekoff, Marc. (2005). *Animal Passions and Beastly Virtues: Reflections on Redecorating Nature*. Philadelphia: Temple University Press.

Bentham, Jeremy. (1789) *An Introduction to the Principles of Morals and Legislation*. London.

Bernard, Claude. (1949) *An Introduction to the Study of Experimental Medicine*. US: Henry Schuman. Originally published in 1865.

Bernstein, Mark. (2004) "Neo-speciesism," *Journal of Social Philosophy*, 35:3, pp. 380–390.

Bonny, Sylvie. (2003) "Why are most Europeans opposed to GMOs? Factors explaining rejection in France and Europe," *Electronic Journal of Biotechnology*, 6:1, pp. 50–71.

Brody, Baruch A. (1988) *Life and Death Decision Making*. Oxford: Oxford University Press.

———. (1998) *The Ethics of Biomedical Research. An International Perspective*. Oxford: Oxford University Press.

———. (2001). "Defending Animal Research: An International Perspective." In *Why Animal Experimentation Matters: The Use of Animals in Medical Research*, eds. Ellen Frankel Paul and Jeffrey Paul. New Brunswick: Transaction Publishers, pp. 131–147.

Broom, Donald M. (1991) "Animal welfare: concepts and measurement," *Journal of Animal Science*, 69, pp. 4167–4175.

———. (1993) "A Usable Definition of Animal Welfare," *Journal of Agricultural and Environmental Ethics*, 6 (Special supplement 2), pp. 15–25.

———. (1996) "Animal welfare defined in terms of attempts to cope with the environment," *Acta Agriculturae Scandinavica* Section A Animal Science (Supplement 27), pp. 22–28.

Bruce, Donald. (2002) "Engineering Genesis: Pioneering Genetic Engineering and Ethics in Scotland." In *Genetic Engineering and the Intrinsic Value and Integrity of Animal and Plants*, eds. David Heaf and Johannes Wirz. Hafan: Ifgene, pp. 11–16.

Buckle, Stephen. (1991) *Natural Law and the Theory of Property: Grotius to Hume*. Oxford: Oxford University Press.

Buehr, Mia, J. Peter Hjort, Axel Kornerup Hansen, and Peter Sandøe. (2003) "Genetically Modified Laboratory Animals: What Welfare Problems Do They Face?," *Journal of Applied Animal Welfare Science*, 6:4, pp. 319–338.

BVAAWF/FRAME/RSPCA/UFAW Joint Working Group on Refinement. (2003) "Refinement and reduction in production of genetically modified mice," *Laboratory Animals*, 37 (Supplement 1), pp. s1–s49.

Byrne, J. A., D. A. Pedersen, L. L. Clepper, M. Nelson, W. G. Sanger, S. Gokhale, D. P. Wolf, and S. M. Milalipov. (2007) "Producing primate embryonic stem cells by somatic cell nuclear transfer," *Nature*, 450:7169, pp. 497–502.

Canadian Council on Animal Care (CCAC). (1997) *The CCAC Guidelines on Transgenic Animals*. Canada.

Capaldi, Nicholas. (1989) *Hume's Place in Moral Philosophy*. New York: Peter Lang.

Capecchi, Mario R. (1989) "The new mouse genetics: altering the genome by gene targeting," *Trends in Genetics*, 5, pp. 70–76.

Carruthers, Peter. (1992) *The Animals Issue: Moral Theory in Practice*. Cambridge: Cambridge University Press.

Chan, A. W. S., K. Y. Chong, C. Martinovich, C. Simerly, and G. Schatten. (2001) "Transgenic monkeys produced by retroviral gene transfer into mature oocytes," *Science*, 291:5502, pp. 309–312.

Chandroo, Kristopher Paul, Stephanie Yue, and Richard David Moccia. (2004) "An evaluation of current perspectives on consciousness and pain in fishes," *Fish and Fisheries*, 5, pp. 281–295.

Cohen, Carl. (1994) "The Case for the Use of Animals in Biomedical Research." In *Ethical Issues in Scientific Research: An Anthology*, eds. Edward Erwin, Sidney Gendin, and Lowell Kleiman. New York: Garland Publishers, pp. 253–266. Originally published in 1986 in the *New England Journal of Medicine*, 315:14, pp. 865–870.

Cohen, Carl, and Tom Regan. (2001) *The Animal Rights Debate*. Lanham: Rowman & Littlefield Publishers.

Commission of the European Communities. (2005) *Fourth Report from the Commission to the Council and the European Parliament on the Statistics on the number of animals used for experimental and other scientific purposes in the member states of the European Union.* http://ec.europa.eu/environment/chemicals/lab_animals/statistics_reports_en.htm (accessed 30 April 2009).

Council Directive 86/609/EEC on the approximation of laws, regulations and administrative provisions of the Member States regarding the protection of animals used for experimental and other scientific purposes. (1986) http://ec.europa.eu/food/fs/aw/aw_legislation/scientific/86-609-eec_en.pdf (accessed 30 April 2009).

Crawley, Jacqueline N. (2000) *What Is Wrong with my Mouse? Behavioural Phenotyping of Transgenic and Knockout Mice.* New York: Wiley-Liss.

Damasio, Antonio R. (1994) *Descartes' Error: Emotion, Reason, and the Human Brain.* New York: G. P. Putnam's Sons.

Daniels, Norman. (1979) "Wide reflective equilibrium and theory acceptance in ethics," *The Journal of Philosophy*, 76, pp. 256–282.

Dawkins, Marian Stamp. (1980) *Animal Suffering: The Science of Animal Welfare.* London: Chapman and Hall.

———. (1988) "Behavioural deprivation: a central problem in animal welfare," *Applied Animal Behaviour Science*, 20, pp. 200–225.

DeGrazia, David. (1996) *Taking Animals Seriously: Mental Life and Moral Status.* Cambridge: Cambridge University Press.

de Lecea, L., T. S. Kilduff, C. Peyron, X.-B. Gao, P. E. Foye, P. E. Danielson, C. Fukuhara, E. L. F. Battenberg, V. T. Gautvik, F. S. Bartlett II, W. N. Frankel, A. N. van den Pol, F. E. Bloom, K. M. Gautvik, and J. G., Sutcliffe. (1998) "The hypocretins: hypothalamus-specific peptides with neuroexcitatory activity," *Proceedings of The National Academy of Sciences of The United States of America*, 95, pp. 322–327.

Descartes, René. (1997) "Discourse on the Method." In *Key Philosophical Writings*, translated by Elizabeth S. Haldane and G. R. T. Ross, and edited and with an Introduction by Enrique Chávez-Arvizo. Hertfordshire: Wordsworth, pp. 71–122. Originally published in 1637.

de Waal, Frans. (1982) *Chimpanzee Politics*. London: Unwin.

Directive 2001/18/EC on the deliberate release into the environment of genetically modified organisms and repealing Council Directive 90/220/EEC. (2001) http://eur-lex.europa.eu/LexUriServ/LexUriServ.do?uri=OJ:L:2001:106:0001:0038:EN:PDF (accessed 30 April 2009)

Doetschman, Thomas. (2002) "Gene Targeting in Embryonic Stem Cells: I. History and Methodology." In *Transgenic Animal Technology: A Laboratory Handbook* (2nd ed.), ed. Carl A. Pinkert. Amsterdam: Academic Press, pp. 113-141.

Dol, Marcel, Martje Fetener Vlissingen, Soemini Kasanmoentalib, Thijs Visser, and Hub Zwart, eds. (1999) *Recognizing the Intrinsic Value of Animals. Beyond Animal Welfare*. Assen: Van Gorcum.

Duncan, Ian J. H. (1993) "Welfare is to do with what animals feel," *Journal of Agricultural and Environmental Ethics*, 6 (Special supplement 2), pp. 8–14.

Duncan, Ian J. H., and David Fraser. (1997) "Understanding animal welfare." In Appleby and Hughes, eds. *Animal Welfare*.

Durant, John, Martin W. Bauer, and George Gaskell, eds. (1998) *Biotechnology in the Public Sphere. A European Sourcebook*. London: The Board of Trustees of the Science Museum.

Erwin, Edward, Sidney Gendin, and Lowell Kleiman, eds. (1994) *Ethical Issues in Scientific Research: An Anthology*. New York: Garland Publishers.

Esbjornson, Robert, ed. (1984) *The Manipulation of Life*. San Francisco: Harper & Row.

Eurobarometer 46.1. (1997) *The Europeans and Modern Biotechnology*. Brussels: European Commission.
http://ec.europa.eu/public_opinion/archives/ebs/ebs_108_en.pdf (accessed 30 April 2009).

Eurobarometer 52.1. (2000) *The Europeans and Biotechnology*. Brussels: European Commission.
http://ec.europa.eu/public_opinion/archives/ebs/ebs_134_en.pdf (accessed 30 April 2009).

Eurobarometer 58.0. (2003) *Europeans and Biotechnology in 2002* (2nd ed.). Brussels: European Commission.
http://ec.europa.eu/public_opinion/archives/ebs/ebs_177_en.pdf (accessed 30 April 2009).

Eurobarometer 64.3. (2006) *Europeans and Biotechnology in 2005: Patterns and Trends*. Brussels: European Commission.
http://ec.europa.eu/public_opinion/archives/ebs/ebs_244b_en.pdf (accessed 30 April 2009).

Farm Animal Welfare Council (FAWC). (2009) *Five Freedoms*.
www.fawc.org.uk/freedoms.htm (accessed 30 April 2009)

Festing, Michael F. (1994) "Reduction of animal use: experimental design and quality of experiments," *Laboratory Animals*, 28, pp. 212–221.

———. (2001) "Sacred cows and golden geese," *Alternatives To Laboratory Animals*, 29, pp. 617–619.

Fletcher, Joseph. (1974) *The Ethics of Genetic Control: Ending Reproductive Roulette*. Garden City: Anchor.

Food Animal Well-Being. (1993) West Lafayette: Purdue University Office of Agricultural Research Programs.

Fox, Michael. (1990) "Transgenic animals: Ethical and animal welfare concerns." In *The Bio-Revolution: Cornucopia or Pandora's Box?*, eds. Peter Wheale and Ruth McNally. London: Pluto, pp 31–45.

Fraser, David. (1993) "Assessing animal well-being: Common sense, uncommon science." In *Food Animal Well-Being*. West Lafayette: Purdue University Office of Agricultural Research Programs, pp. 37-57.

———. (1995) "Science, values and animal welfare: exploring the 'inextricable connection'," *Animal Welfare*, 4, pp. 103–117.

———. (1999) "Animal ethics and animal welfare science: bridging the two cultures," *Applied Animal Behaviour Science*, 65, pp. 171–189.

———. (2003) "Assessing animal welfare at the farm and group level: the interplay of science and values," *Animal Welfare*, 12, pp. 433–443.

Fraser, D., D. M. Weary, E. A. Pajor, and B. N. Milligan. (1997) "A scientific conception of animal welfare that reflects ethical concerns," *Animal Welfare*, 6, pp. 187–205.

Frey, R. G. (1980) *Interests and Rights: The Case Against Animals*. Oxford: Clarendon.

Fulka Jr, Josef, Norikazu Miyashita, Takashi Nagai, and Atsuo Ogura. (2004) "Do cloned mammals skip a reprogramming step?," *Nature Biotechnology*, 22, pp. 25–26.

Gibson, F., J. Walsh, P. Mburu, A. Varela, K. A. Brown, M. Antonio, K. W. Beisel, K. P. Steel, and S. D. M. Brown. (1995) "A type VII myosin encoded by the mouse deafness gene shaker-1," *Nature*, 374, pp. 62–64.

Gordon, Jon W., George A. Scangos, Diane J. Plotkin, James A. Barbosa, and Frank H. Ruddle. (1980) "Genetic transformation of mouse embryos by microinjection of purified DNA," *Proceedings of The National Academy of Sciences of The United States of America*, 77, pp. 7380–7384.

Greek, C. Ray, and Jean Swingle Greek. (2001) *Sacred Cows and Golden Geese: The Human Cost of Experiments on Animals*. New York: Continuum.

———. (2002) *Specious Science: How Genetics and Evolution Reveal Why Medical Resesarch on Animals Harms Humans*. New York: Continuum.

Guerrini, Anita. (2004) Review (of Greek and Greek (2002)), *Journal of the History of Medicine*, 59, pp. 164–165.

Hare, Richard M. (1981) *Moral Thinking: Its Levels, Method, and Point*. Oxford: Clarendon Press.

Heaf, David, and Johannes Wirz, eds. (2002) *Genetic Engineering and the Intrinsic Value and Integrity of Animal and Plants*. Hafan: Ifgene.

Heeger, Robert. (1997) "Respect for Animal Integrity?" In *Science, Ethics, Sustainability: The Responsibility of Science in Attaining Sustainable Development*, ed. Anders Nordgren. Studies in Bioethics and Research Ethics 2. Uppsala: Acta Universitatis Uppsaliensis, pp. 243–252.

Hochedlinger, Konrad, and Rudolf Jaenisch. (2003) "Nuclear Transplantation, Embryonic Stem Cells, and the Potential for Cell Therapy," *The New England Journal of Medicine*, 349:3, pp. 275–286.

Houdebine, Louis-Marie. (2003) *Animal Transgenesis and Cloning*. Chichester: John Wiley.

Hull, David L., and Michael Ruse, eds. (1998) *The Philosophy of Biology*. Oxford: Oxford University Press.

Hume, David. (1978) *A Treatise of Human Nature*. Oxford: Oxford University Press. Originally published in 1739–1740.

Humpherys, David, Kevin Eggan, Hidenori Akutsu, Adam Friedman, Konrad Hochedlinger, Ryuzo Yanagimachi, Eric S. Lander, Todd R. Golub, and Rudolf Jaenisch. (2002) "Abnormal gene expression in cloned mice derived from embryonic stem cell and cumulus cell nuclei," *Proceedings of the National Academy of Sciences of the United States of America*, 99, pp. 12889–12894.

Hursthouse, Rosalind. (2000) *Ethics, Humans and Other Animals*. London: Routledge.

International Human Genome Sequencing Consortium. (2001) "Initial sequencing and analysis of the human genome," *Nature*, 409, pp. 860–921.

Jaenisch, Rudolf, and Adrian Bird. (2003) "Epigenetic regulation of gene expression: how the genome integrates intrinsic and environmental signals," *Nature Genetics*, 33 (suppl.), pp. 245–254.

Jaenisch, Rudolf, and Ian Wilmut. (2001) "Developmental Biology: Don't Clone Humans!," *Science*, 291, p. 2552.

Jaffé, Walter, and Miguel Rojas. (1994) "Transgenic Potato Tolerant to Freezing," *Biotechnology and Development Monitor*, 18, p. 10.

Jegstrup, Inger, Rikke Thon, Axel Kornerup Hansen, and Merel Ritskes Hoitinga. (2003) "Characterization of transgenic mice: A comparison of protocols for welfare evaluation and phenotype characterization of mice with a suggestion on a future certificate of instruction," *Laboratory Animals*, 37, pp. 1–9.

Johnson, Mark. (1993) *Moral Imagination: Implications of Cognitive Science for Ethics*. Chicago: The University of Chicago Press.

Jonas, Hans. (1984) *The Imperative of Responsibility: In Search of an Ethics for the Technological Age*. Chicago: The University of Chicago Press.

Jonsen, Albert R., and Stephen Toulmin. (1988) *The Abuse of Casuistry: A History of Moral Reasoning*. Berkeley: University of California Press.

Kant, Immanuel. (1963) *Lectures on Ethics (1780-81)*. Translated by H. Louis Infield. New York: Harper and Row.

Kauffman, Stuart A. (1993) *The Origins of Order: Self-Organization and Selection in Evolution*. Oxford: Oxford University Press.

Kiley-Worthington, Marthe. (1989) "Ecological, ethological, and ethically sound environments for animals: toward symbiosis," *Journal of Agricultural Ethics*, 2, pp. 323–347.

Kim, K.-S., J. Kim, S. K. Back, J.-Y. Im, H. S. Na, and P.-L. Han. (2007) "Markedly attenuated acute and chronic pain responses in mice lacking adenylyl cyclase-5," *Genes, Brain and Behavior*, 6:2, pp. 120–127.

Kim, Myoung Ok, Sung Hyun Kim, Mi Jung Shin, Dong Beom Lee, Tae Won Kim, Kil Soo Kim, Ji Hong Ha, Sanggyu Lee, Yong Bok Park, Sun Jung Kim, and Zae Young Ryoo. (2007) "Human erythropoietin Induces Lung Failure and Erythrocytosis in Transgenic Mice," *Molecules and Cells*, 23:1, pp. 17–22.

Kim, Teoan. (2002) Retrovirus-Mediated Gene Transfer. In *Transgenic Animal Technology: A Laboratory Handbook* (2nd ed.), ed. Carl A. Pinkert. Amsterdam: Academic Press, pp. 173–193.

Kumar, R., A. M. Lozano, Y. J. Kim, W. D. Hutchison, E. Sime, E. Halket, and A. E. Lang. (1998) "Double-blind evaluation of subthalamic nucleus deep brain stimulation in advanced Parkinson's disease," *Neurology*, 51, pp. 850–855.

LaFollette, Hugh. (1993) "Personal Relationships." In *A Companion to Ethics*, ed. Peter Singer. Oxford: Blackwell, pp. 327–332.

LaFollette, Hugh, and Niall Shanks. (1996) *Brute Science: Dilemmas of animal experimentation.* London: Routledge.

Lakoff, George, and Mark Johnson. (1980) *Metaphors We Live By.* Chicago: University of Chicago Press.

———. (1999) *Philosophy in the Flesh: The Embodied Mind and Its Challenge to Western Thought.* New York: Basic Books.

Lassen, Jesper, Mickey Gjerris, and Peter Sandøe. (2006) "After Dolly: Ethical limits to the use of biotechnology on farm animals," *Theriogenology*, 65, pp. 992–1004.

Lebacqz, Karen. (1984) "The Ghosts Are on the Wall: A Parable for Manipulating Life." In Esbjornson, ed. *The Manipulation of Life.*

Leahy, Michael P. T. (1991). *Against Liberation: Putting Animals in Perspective.* London: Routledge.

Lee, Byeong Chun, Min Kyu Kim, Goo Jang, Hyun Ju Oh, Fibrianto Yuda, Hye Jin Kim, M. Hossein Shamim, Jung Ju Kim, Sung Keun Kang, Gerald Schatten, and Woo Suk Hwang. (2005) "Dogs cloned from adult somatic cells," *Nature*, 436:4, p. 641.

Lin, Ling, Juliette Faraco, Robin Li, Hiroshi Kadotani, William Rogers, Xiaoyan Lin, Xiaohong Qiu, Pieter J. de Jong, Seiji Nishino, and Emmanuel Mignot. (1999) "The sleep disorder canine narcolepsy is caused by a mutation in the hypocretin (orexin) receptor 2 gene," *Cell*, 98, pp. 365–376.

Linzey, Andrew. (1995) *Animal Theology.* Champaign: University of Illinois Press.

Macnaghten, Phil. (2001) *Animal Futures: Public Attitudes and Sensibilities towards Animals and Biotechnology in Contemporary Britain.* Lancaster: IEPPP. http://www.aebc.gov.uk/aebc/pdf/macnaghten_animals_futures.pdf (accessed 30 April 2009)

———. (2004) "Animals in Their Nature: A Case Study on Public Attitudes to Animals, Genetic Modification and 'Nature'," *Sociology*, 38:3, pp. 533–551.

Mepham, T. Ben, Robert D. Combes, Michael Balls, Ottavia Barbieri, Harry J. Blokhuis, Patrizia Costa, Robert E. Crilly, Tjard de Cock Buning, Véronique C. Delpire, Michael J. O'Hare, Louis-Marie Houdebine, Coen F. van Kreijl, Miriam van der Meer, Christoph A. Reinhardt, Eckhard Wolfe, and Anne-Marie van Zeller. (1998) "The Use of Transgenic Animals in the European Union: The Report and Recommendations of ECVAM Workshop 28," *Alternatives To Laboratory Animals*, 26, pp. 21–43.

Midgley, Mary. (1979) *Beast and Man: The Roots of Human Nature.* London: Methuen.

———. (1981) *Heart and Mind: The Varieties of Moral Experience.* New York: St. Martin's Press.

———. (1983) *Animals and Why They Matter.* Athens: The University of Georgia Press.

———. (2000) "Biotechnology and Monstrosity: Why We Should Pay Attention to the 'Yuk Factor'," *Hastings Center Report*, 30:5 (Sept.–Oct.), pp. 7–15.

Miller, Jon D. (1998) "The measurement of scientific literacy," *Public Understanding of Science*, 7, pp. 203–223.

Mishler, Brent D., and Robert N. Brandon. (1998) "Individuality, Pluralism and the Phylogenetic Species Concept." In *The Philosophy of Biology*, eds. David L. Hull and Michael Ruse. Oxford: Oxford University Press, pp. 300–318.

Moore, Colin J., and T. Ben Mepham. (1995) "Transgenesis and Animal Welfare," *Alternatives To Laboratory Animals*, 23, pp. 380–397.

MORI. (1999) *Animals in Medicine and Science: General Public Research Conducted for Medical Research Council*. London: MRC.

Mouse Genome Sequencing Consortium. (2002) "Initial sequencing and comparative analysis of the mouse genome," *Nature*, 420:6915, pp. 520–562.

Musschenga, Albert W. (2005) "The Debate on Impartiality: An Introduction," *Ethical Theory and Moral Practice*, 8, pp. 1–10.

Nakao, Harumi, Kazuki Nakao, Masanobu Kano, and Atsu Aiba. (2007) "Metabotropic glutamate receptor subtype-1 is essential for motor coordination in the adult cerebellum," *Neuroscience Research*, 57:4, pp. 538–543.

National Academy of Sciences. (2002) *Scientific and Medical Aspects of Human Reproductive Cloning*. Washington D.C.: National Academy Press.

Noddings, Nel. (1984) *Caring: A Feminine Approach to Ethics and Moral Education*. Berkeley: University of California Press.

Nordenfelt, Lennart. (2006) *Animal and Human Health and Welfare: A Comparative Philosophical Analysis*. Wallingford: CABI.

Nordgren, Anders. (1994) *Evolutionary Thinking: An Analysis of Rationality, Morality and Religion from an Evolutionary Perspective*. Stockholm: Almqvist & Wiksell International.

———, ed. (1997) *Science, Ethics, Sustainability: The Responsibility of Science in Attaining Sustainable Development*. Studies in Bioethics and Research Ethics 2. Uppsala: Acta Universitatis Upsaliensis.

———. (1998) "Ethics and Imagination: Implications of Cognitive Semantics for Medical Ethics," *Theoretical Medicine and Bioethics*, 19, pp. 117–141.

———. (2001) *Responsible Genetics: The Moral Responsibility of Geneticists for the Consequences of Human Genetics Research*. Philosophy and Medicine series Vol. 70. Dordrecht: Kluwer Academic Publishers.

———. (2002) "Animal experimentation: pro and con arguments using the theory of evolution," *Medicine, Health Care and Philosophy*, 5, pp. 23–31.

———. (2003) "Metaphors in Behavioral Genetics," *Theoretical Medicine and Bioethics*, 24:1, pp. 59–77.

———. (2004) "Moral imagination in tissue engineering research on animal models," *Biomaterials*, 25, pp. 1723–1734.

———. (2006) "Analysis of an epigenetic argument against human reproductive cloning," *Reproductive BioMedicine Online*, 13:2, pp. 278–283.

Nordgren, A., and H. Röcklinsberg. (2005) "Genetically modified animals in research: An analysis of applications submitted to ethics committees on animal experimentation in Sweden," *Animal Welfare*, 14, pp. 239–248.

Nuffield Council on Bioethics. (2005) *The Ethics of Research Involving Animals.* www.nuffieldbioethics.org (accessed 30 April 2009).

Nussbaum, Martha C. (2003) *Upheavals of Thought: The Intelligence of Emotions* (new ed.) Cambridge: Cambridge University Press.

———. (2006) *Frontiers of Justice: Disability, Nationality, Species Membership.* Cambridge, Mass: The Belknap Press of Harvard University Press.

Orlans, F. Barbara. (1993) *In the Name of Science: Issues in Responsible Animal Experimentation.* New York: Oxford University Press.

Parens, Erik, Audrey R. Chapman, and Nancy Press, eds. (2006) *Wrestling with Behavioral Genetics: Science, Ethics, and Public Conversation.* Baltimore: The Johns Hopkins University Press.

Paterson, Lesley, William Ritchie, and Ian Wilmut. 2002. "Nuclear Transfer Technology in Cattle, Sheep, and Swine." In *Transgenic Animal Technology: A Laboratory Handbook* (2nd ed.), ed. Carl A. Pinkert. Amsterdam: Academic Press, pp. 395–416.

Paul, Ellen Frankel, and Jeffrey Paul, eds. (2001) *Why Animal Experimentation Matters: The Use of Animals in Medical Research.* New Brunswick: Transaction Publishers.

Peters, Ted (1997) *Playing God? Genetic Determinism and Human Freedom.* New York: Routledge.

Petrinovich, Lewis (1998) *Human Evolution, Reproduction, and Morality.* Cambridge: The MIT Press. Originally published in 1995 by Plenum Press, New York.

———. (1999) *Darwinian Dominion: Animal Welfare and Human Interests.* Cambridge: The MIT Press.

Peyron, Christelle, Juliette Faraco, William Rogers, Beth Ripley, Sebastiaan Overeem, Yves Charnay, Sona Nevsimalova, Michael Aldrich, *et al.* (2000) "A mutation in a case of early onset narcolepsy and a generalized absence of hypocretin peptides in human narcoleptic brains," *Nature Medicine*, 6, pp. 991–997.

Pinkert, Carl A., ed. (2002) *Transgenic Animal Technology: A Laboratory Handbook* (2nd ed.). Amsterdam: Academic Press.

Polites, H. G., and Carl A. Pinkert. (2002) "DNA Microinjection and Transgenic Animal Production." In *Transgenic Animal Technology: A Laboratory Handbook* (2nd ed.), ed. Carl A. Pinkert. Amsterdam: Academic Press, pp. 15–70.

Post, Stephen G. (1993) "The Emergence of Species Impartiality: A Medical Critique of Biocentrism," *Perspectives in Biology and Medicine*, 36, pp. 289–300.

President's Commission for the Study of Ethical Problems in Medicine and Biomedical and Behavioral Research. (1982) *Splicing Life: A Report on the Social and Ethical Issues of Genetic Engineering with Human Beings.* Washington D.C.: US Government Printing Office.

Public Health Service. (1986) *Principles for the Utilization and Care of Vertebrate Animals Used in Testing, Research, and Training.* U.S. Government.

———. (2002) *Policy on Humane Care and Use of Laboratory Animals.* USA.

Radcliffe Richards, Janet. (2000) *Human Nature after Darwin: A Philosophical Introduction.* London: Routledge.

Ramsey, Paul. (1970) *The Patient as Person: Explorations in Medical Ethics.* New Haven: Yale University Press.

Rat Genome Sequencing Project Connsortium. (2004) "Genome sequence of the Brown Norway rat yields insights into mammalian evolution," *Nature*, 428, pp. 493–521.
Rawls, John. (1971) *A Theory of Justice*. Oxford: Oxford University Press.
———. (1993) *Political Liberalism*. New York: Columbia University Press.
Regan, Tom. (1983) *The Case for Animal Rights*. Berkeley: University of California Press.
Reiss, Michael J., and Roger Straughan. (1996) *Improving Nature? The science and ethics of genetic engineering*. Cambridge: Cambridge University Press.
Rideout, William M., Kevin Eggan, and Rudolf Jaenisch. (2001) "Nuclear cloning and epigenetic reprogramming of the genome," *Science*, 293, pp. 1093–1098.
Rodd, Rosemary. (1990) *Biology, Ethics, and Animals*. Oxford: Oxford University Press.
Rodriguez-Oroz, M. C., I. Zamarbide, J. Guridi, M. R. Palmera, and J. A. Obeso. (2004) "Efficacy of deep brain stiumlation of the subthalamic nucleus in Parkinson's disease four years after surgery. Double blind and open label evaluation," *Journal of Neurology, Neurosurgery and Psychiatry*, 75, pp. 1382–1385.
Rollin, Bernard E. (1981) *Animal Rights and Human Morality*. Buffalo: Prometheus.
———. (1989) *The Unheeded Cry: Animal Consciousness, Animal Pain and Science*. Oxford: Oxford University Press.
———. (1993) "Animal welfare, science and value," *Journal of Agricultural and Environmental Ethics*, 6 (Suppl 2), pp. 44–50.
———. (1995) *The Frankenstein Syndrome: Ethical and Social Issues in the Genetic Engineering of Animals*. Cambridge: Cambridge University Press.
Rolston III, Holmes. (2002) "What do we mean by the intrinsic value and integrity of plants and animals?" In *Genetic Engineering and the Intrinsic Value and Integrity of Animal and Plants*, eds. David Heaf and Johannes Wirz. Hafan: Ifgene, pp. 5–10.
Rosch, Eleanor, and Barbara L. Lloyd, eds. (1978) *Cognition and Categorization*. Hillsdale: Lawrence Erlbaum.
Rose, James D. (2002) "The neurobehavioural nature of fishes and the question of awareness and pain," *Reviews in Fisheries Science*, 10, pp. 1–38.
Rowan, Andrew N. (1984) *Of Mice, Models, and Men*. New York: State University of New York.
Royal Society. (2001) *The Use of Genetically Modified Animals*. Policy document 5/01, The United Kingdom.
Rucker III, Edmund B., James G. Thomson, and Jorge A. Piedrahita. (2002) "Gene Targeting in Embryonic Stem Cells: II. Conditional Technologies." In *Transgenic Animal Technology: A Laboratory Handbook* (2nd ed.), ed. Carl A. Pinkert. Amsterdam: Academic Press, pp. 143–171.
Russell, W. M. S., and R. L. Burch. (1992) *The Principles of Humane Experimental Technique*. Wheathampstead: Universities Federation for Animal Welfare. Originally published in 1959, London: Methuen.
Rutgers, Bart, and Robert Heeger. (1999) "Inherent Worth and Respect for Animal Integrity." In *Recognizing the Intrinsic Value of Animals. Beyond Animal Wel-

fare, eds. Marcel Dol, Martje Fetener Vlissingen, Soemini Kasanmoentalib, Thijs Visser, and Hub Zwart. Assen: Van Gorcum, pp. 41–51.

Ryder, Richard D. (1999) "Painism: some moral rules for the civilized experimenter," *Cambridge Quarterly of Healthcare Ethics*, 8, pp. 35–42.

Sakurai, Takeshi, Akira Amemiya, Makoto Ishii, Ichiyo Matsuzaki, Richard M. Chemelli, Hirokazu Tanaka, S. Clay Williams, James A. Richardson, Gerald P. Kozlowski, Shelagh Wilson, Jonathan R. S. Arch, Robin E. Buckingham, Andrea C. Haynes, Steven A. Carr, Roland S. Annan, Dean E. McNulty, Wu-Schyong Liu, Jonathan A. Terrett, Nabil A. Elshourbagy, Jerk J. Bergsma, and Masashi Yanagisawa. (1998) "Orexins and orexin receptors: a family of hypothalamic neuropeptides and G protein-coupled receptors that regulate feeding behaviour," *Cell*, 92, pp. 573–585.

Sandøe, Peter. (1996) "Animal and human welfare: are they the same kind of thing?," *Acta Agriculturae Scandinavica* Section A Animal Science (Supplement 27), pp. 11–15.

Sapontzis, Steve F. (1987) *Morals, Reason, and Animals*. Philadelphia: Temple.

Savage-Rumbaugh, Sue, Stuart G. Shanker, and Talbot J. Taylor. (1998) *Apes, Language, and the Human Mind*. Oxford: Oxford University Press.

Segerdahl, Pär, William Fields, and Sue Savage-Rumbaugh. (2006) *Kanzi's Primal Language: The Cultural Initiation of Primates into Language*. Basingstoke: Palgrave Macmillan.

Sherlock, Richard. (2002) "Three Concepts of Genetic Trespassing." In *Ethical Issues in Biotechnology*, eds. Richard Sherlock and John D. Morrey. Lanham: Rowman & Littlefield Publishers, pp. 149–159.

Sherlock, Richard, and John D. Morrey, eds. (2002) *Ethical Issues in Biotechnology*. Lanham: Rowman & Littlefield Publishers.

Sigmund, Curt D. (2000) "Viewpoint: Are Studies in Genetically Altered Mice Out of Control?," *Arteriosclerosis, Thrombosis, and Vascular Biology*, 20, p. 1425.

Simpson, Joe Leigh. (2003) "Toward scientific discussion on human reproductive cloning," *Reproductive BioMedicine Online*, 7, pp. 10–11.

Singer, Peter. (1993a) *Practical Ethics* (2nd ed.; 1st ed. 1979). Cambridge: Cambridge University Press.

———, ed. (1993b) *A Companion to Ethics*. Oxford: Blackwell.

———. (1994) "The Significance of Animal Suffering." In *Ethical Issues in Scientific Research: An Anthology*, eds. Edward Erwin, Sidney Gendin, and Lowell Kleiman. New York: Garland Publishers, pp. 233–243.

———. (1995)) *Animal liberation* (2nd ed. with a new preface by the author; 1st ed. 1975). London: Pimlico.

Sjöberg, Lennart. (2004) *Gene Technology in the Eyes of the Public and Experts: Moral Opinions, Attitudes and Risk Perception*. SSE/EFI Working Paper Series in Business Administration No 2004:7. Center of Risk Research. Stockholm School of Business. Stockholm.

Smith, Jane A., and Kenneth M. Boyd. (1991) *Lives in the Balance: The Ethics of Using Animals in Biomedical Research*. Oxford: Oxford University Press.

Smith, Richard. (2001) "Animal research: the need for a middle ground," *British Medical Journal*, 322, pp. 248–249.

Stafleu, F. R., F. J. Grommers, and J. Vorstenbosch. (1996) "Animal welfare: evolution and erosion of a moral concept," *Animal Welfare*, 5, pp. 225–234.

Stafleu, F. R., R. Tramper, J. Vorstenbosch, and J. A. Joles. (1999) "The ethical acceptability of animal experiments: a proposal for a system to support decision-making," *Laboratory Animals*, 33, pp. 295–303.

Steel, Daniel P. (2008) *Across the Boundaries: Extrapolation in Biology and Social Science*. Oxford: Oxford University Press.

Stokstad, Erik. (1999) "Humane science finds sharper and kinder tools," *Science*, 286, pp. 1068–1071.

Strong, Carson. (1997) *Ethics in Reproductive and Perinatal Medicine: A New Framework*. New Haven: Yale University Press.

Sumner, L. Wayne. (1996) *Welfare, Happiness, and Ethics*. Oxford: Clarendon Press.

Tannenbaum, Jerrold. (1991) "Ethics and animal welfare: The inextricable connection," *Journal of American Veterinary Medical Association*, 198:8, pp. 1360–1376.

———. (2001) "The Paradigm Shift toward Animal Happiness: What It Is, Why It Is Happening, and What It Portends for Medical Research." In *Why Animal Experimentation Matters: The Use of Animals in Medical Research*, eds. Ellen Frankel Paul and Jeffrey Paul. New Brunswick: Transaction Publishers, pp. 93-130.

Thannickal, Thomas C., Robert Y. Moore, Robert Nienhuis, Lalini Ramanathan, Seema Gulyani, Michael Aldrich, Marsha Cornford, and Jerome M. Siegel. (2000) "Reduced number of hypocretin neurons in human narcolepsy," *Neuron*, 27, pp. 469–474.

The Chimpanzee Sequencing and Analysis Consortium. (2005) "Initial sequence of the chimpanzee genome and comparison with the human genome," *Nature*, 437, pp. 69-87.

Thon, Rikke, Jesper Lassen, Axel Kornerup Hansen, Inger Marie Jegstrup, and Merel Ritskes-Hoitinga. (2002) "Welfare evaluation of genetically modified mice: An inventory study of reports to the Danish Animal Experiments Inspectorate," *Scandinavian Journal of Laboratory Animal Science*, 1, pp. 45–53.

Toulmin, Stephen. (1958) *The Uses of Argument*. Cambridge: Cambridge University Press.

Toye, Ayo A., Lee Moir, Alison Hugill, Liz Bentley, Julie Quarterman, Vesna Mijat, Tertius Hough, Michelle Goldsworthy, Alison Haynes, A. Jacqueline Hunter, Mick Browne, Nigel Spurr, and Roger D. Cox. (2004) "A New Mouse Model of Type 2 Diabetes, Produced by N-Ethyl-Nitrosourea Mutagenesis, is the Result of a Missense Mutation in the Glucokinase Gene," *Diabetes*, 53, pp. 1577–1583.

Tsunoda, Yukio, and Yoko Kato. (2002) "Nuclear Transfer Technologies." In *Transgenic Animal Technology: A Laboratory Handbook* (2nd ed.), ed. Carl A. Pinkert. Amsterdam: Academic Press, pp. 195–231.

UNESCO. (2006) *Universal Declaration on Bioethics and Human Rights*. http://unesdoc.unesco.org/ulis/index.shtml (accessed 30 April 2009).

van der Meer, Miriam. (2001) *Transgenesis and Animal Welfare: Implications of Transgenic Procedures for the Well-Being of the Laboratory Mouse*. Utrecht: Labor Grafimedia BV.

van der Meer, M., A. Rolls, V. Baumans, B. Olivier, and L. F. M. van Zutphen. (2001) "Use of score sheets for welfare assessment of transgenic mice," *Laboratory Animals*, 35, pp. 379–389.
van Zutphen, B., and M. van der Meer, eds. (1997) *Welfare of Transgenic Animals.* Berlin: Springer Verlag.
Venter, J. Craig, Mark D. Adams, Eugene W. Myers, Peter W. Li, Richard J. Mural, Granger G. Sutton, Hamilton O. Smith, Mark Yandell, *et al.* (2001) "The Sequence of the Human Genome," *Science*, 291, pp. 1304–1351.
Verhey, Allen. (2002) "'Playing God' and Invoking a Perspective." In *Ethical Issues in Biotechnology*, eds. Richard Sherlock and John D. Morrey. Lanham: Rowman & Littlefield Publishers, pp. 71–87.
Verhoog, Henk. (1992) "The concept of intrinsic value and transgenic animals," *Journal of Agricultural and Environmental Ethics*, 2, pp. 147–160.
Vilcek, Jan, and Marc Feldmann. (2004) "Historical review: cytokines as theprapeutics and targets of therapeutics," *Trends in Pharmacological Sciences*, 25, pp. 201–209.
Warren, Mary Anne. (1997) *Moral Status: Obligations to Persons and Other Living Things.* Oxford: Oxford University Press.
Welchman, Jennifer. (2003) "Xenografting, Species Loyalty, and Human Solidarity," *Journal of Social Philosophy*, 34:2, pp. 244–255.
Wheale, Peter, and Ruth McNally, eds. (1990) *The Bio-Revolution: Cornucopia or Pandora's Box?* London: Pluto.
Wilmut, I., A. E. Schnieke, J. McWhir, A. J. Kind, and K. H. S. Campbell. (1997) "Viable offspring derived from fetal and adult mammalian cells," *Nature*, 385, pp. 810–813.
Wilson, James Q. (1993) *The Moral Sense.* New York: Free Press.
Winkler, Earl R. (1993) "From Kantianism to Contextualism: The Rise and Fall of the Paradigm Theory in Bioethics." In *Applied Ethics: A Reader*, eds. Earl R. Winkler and Jerrold R. Coombs. Oxford: Blackwell, pp. 343–365.
Winkler, Earl R., and Jerrold R. Coombs, eds. (1993) *Applied Ethics: A Reader.* Oxford: Blackwell.
World Medical Association. (1964) (with later revisions). *Declaration of Helsinki.* www.wma.net/e/policy/b3.htm (accessed 30 April 2009).
Zambrowicz, Brian P., and Arthur T. Sands. (2003) "Knockouts model the 100 best-selling drugs: will they model the next 100?," *Nature Reviews Drug Discovery*, 2, pp. 38–51.
Zamir, Tzachi. (2006) "Killing for Knowledge," *Journal of Applied Philosophy*, 23:1, pp. 17–40.
———. (2007) *Ethics and the Beast: A Speciesist Argument for Animal Liberation.* Princeton: Princeton University Press.

INDEX

aggregation, 107, 138–141, 143, 144
animal experiment(s)(ation), 1–4, 6–14, 17–18, 20–24, 26, 29, 34–36, 39–40, 42, 44–45, 47–52, 61, 62, 63, 67–70, 72–78, 82–83, 85–86, 90–92, 94–95, 97, 98, 99, 100, 101–102, 104, 106, 107, 108, 111, 120, 123, 125, 127, 129, 130, 134–135, 137–142, 143, 144, 145, 147–148, 150, 154, 155, 164, 165, 167, 168, 170, 172, 173, 174, 177, 179
animal integrity, 3, 10, 69–70, 83, 137, 139, 143, 145, 155, 157, 163–166, 175, 177
animal rights, 13, 29, 32, 35–36, 45, 49, 52, 53, 59, 68, 104, 122, 125, 147–148, 156
animal telos, 114, 157, 161–164
animal welfare, 3, 8, 10, 11, 12, 39, 42, 45, 47, 48, 75, 76, 77, 78, 83, 102, 103, 104, 107, 108, 111–128, 134, 136, 137, 139, 143, 148, 162–163, 165–170, 173, 175, 178
antispeciesism, 27–28, 59
Arnhart, Larry, 64, 65
Augustine, 18

balancing, 11, 12, 15, 33, 36, 48, 49, 51, 52, 58, 65, 66, 67, 70, 74, 75, 76, 77, 78, 79, 80, 81, 83, 107, 111, 120, 124, 125, 126, 135, 137–140, 142, 143,–144, 156, 164, 165, 166
Barnard, C. J., 112, 129
Bateson, Patrick, 142
Beauchamp, Tom L., 79
Bentham, Jeremy, 21, 131, 140–141
Bernard, Claude, 20, 96, 97
Bernstein, Mark, 73
Boyd Group, 91, 138, 142
Brody, Baruch A., 13, 49, 55, 60, 61, 72–73, 79, 124
Broom, Donald M., 111–112, 116, 122

Bruce, Donald, 164
Burch, R. L., 12, 63, 75, 103, 107, 108, 122

Canada, 47
care, 15, 17, 20, 41, 43, 48, 50, 61, 62, 63, 65, 67, 72, 73, 74, 80, 108, 122, 126, 143, 144, 156, 169, 172
Carruthers, Peter, 11, 13, 14–18, 19, 30, 37, 39, 50, 51, 52, 53, 54, 56, 59, 63, 68, 69, 104, 122, 125, 127, 128, 132
casuistry, 122, 124, 145
    imaginative c., 58, 77, 78, 79–82, 83, 124, 137, 144
Childress, James F., 79
cloning, 6–7, 9, 87, 149–150, 153, 164–165, 167, 170–171, 177–179
Cohen, Carl, 11, 13, 35–39, 40, 44, 50, 51, 53, 54, 56, 59, 63, 68, 69, 92, 104, 107, 122, 125, 128, 147
conditional methods, 5, 7, 152, 169, 173
contractualism, 14, 15, 16

Damasio, Antonio, 56–57
Daniels, Norman, 79
Dawkins, Marian S., 113
Dawkins, Richard, 102
DeGrazia, David, 11, 14, 16, 42, 128–129, 130–134
Descartes, René, 18–19, 56, 132
distress, 2, 48, 103, 104, 108, 122, 124, 128, 130–131, 136, 137, 166, 167, 170, 177
Duncan, Ian J. H., 111, 113, 122

embryonic stem cell method, 5, 150, 151, 153, 160, 166, 169
epigenetics, 102, 170
equal consideration, 13, 15, 21, 22, 25, 28, 45, 49, 51, 52, 54, 55, 61, 104, 125, 126, 139, 144, 147
Eurobarometer, 9, 10, 155, 156

European Union, 1, 2, 3, 4, 5, 9, 48, 76, 85, 104, 105, 178
evolution, 41, 55, 60–67, 83, 94, 95–97, 101, 102, 113, 129, 133, 160

feeling, 10, 12, 15, 17, 25, 32, 49, 55–58, 59, 60, 63, 64, 65, 67, 68, 77, 83, 106, 107, 112, 113, 114, 116, 117, 118, 119, 120, 121, 122, 123, 124, 125, 126, 127, 130, 131, 134, 136, 137, 139, 143, 147, 156–157, 163, 165, 166, 167, 169, 173, 175
Festing, Michael F., 98, 107
Fields, William, 132
Fletcher, Joseph, 158
Fox, Micheal, 161, 163, 164
France, 1, 3, 47, 85
Fraser, Davis, 11, 12, 111, 116, 117–118, 119–121, 122–123, 124, 126–127, 134, 139
function(ing), 5, 8, 12, 14, 34, 57, 67, 77, 83, 85, 86, 87, 88, 94, 96–97, 98, 111–112, 116, 117, 118, 119, 120, 121, 122, 123, 124, 127, 134, 135, 136, 137, 139, 143, 144, 145, 148, 149, 154, 163, 164, 165, 166, 168, 169, 175, 176

gene regulation, 97, 102
genetically modified animal, 1, 2, 4–5, 7, 9, 10, 12, 86, 88, 89, 99, 147, 148–149, 150, 153, 154, 155, 157, 159, 160, 165–169, 171–173, 179
Germany, 47
Gould, Stephen Jay, 102
Greek, C. Ray, 11, 14, 92, 98, 100
Greek, Jean Swingle, 11, 14, 92, 98, 100

Hare, Richard, 58
Heeger, Robert, 69–70, 142, 163
human dominion, 13, 14, 15, 17, 18, 20, 21, 45, 49, 51, 52, 54, 104, 125, 127, 147, 148
Hume, David, 56, 59, 60, 63–64, 161
Hurst, J. L., 112, 129

Hursthouse, Rosalind, 19, 23–24, 28, 42, 60

impartiality, 24, 49, 58, 59, 60, 61, 63, 68, 70, 71
intrinsic properties, 55, 63, 68

Johnson, Mark, 52, 58, 78, 79, 81, 119
Jonas, Hans, 72
justice, 29, 31, 32, 41, 43, 44, 50, 51, 56, 60, 65, 67, 70, 71, 80, 114, 115, 138
   interspecies j., 14, 32, 43, 44, 50, 54, 60, 62, 67, 70, 74, 76, 83

Kant, Immanuel, 17, 19, 32, 56, 59, 64
Kauffman, Stuart, 97, 102
Kiley-Worthington, Marthe, 113
knock-out, 4, 6, 7, 99, 121, 143, 147, 152, 154, 166, 168, 172, 175–177

LaFollette, Hugh, 1, 11, 14, 59, 60, 91–103
Lakoff, George, 52, 78, 79, 119
Leahy, Michael P. T., 42
loyalty, 71–72

Macnaghten, Phil, 8, 9, 10, 156
Mayr, Ernst, 102
Mepham, T. Ben, 5, 166, 167, 170, 173
metaphor, 14, 18, 30, 49, 52–54, 69, 78, 79, 80, 82, 119, 140, 157, 160, 161
Midgley, Mary, 11, 13, 14, 27, 39–44, 50, 51, 53, 54, 55, 56, 58, 59–61, 62, 63, 64–65, 66–67, 68–69, 70, 73, 75, 76–78, 122, 124, 125, 126, 127, 128, 147, 156, 157, 158, 159
Moore, Colin J., 166, 167, 170, 173
moral imagination, 52, 58, 77, 78, 79, 80, 81, 82–83, 130, 157
Musschenga, Albert W., 62

natural behavior, 116, 123, 162
natural living, 12, 78, 83, 111, 113, 115, 116, 117, 118, 119, 120, 121, 122, 123, 124, 134, 136,

139, 143, 162, 165, 166, 167, 173, 175
natural order, 2, 156, 157, 159, 160–161, 164, 165
nature, 10, 14, 41, 66, 88, 115, 116, 159, 160–161, 173
  animal n., 114, 116–117, 162, 163, 168, 173
  human n., 58, 64, 65–66, 67–68, 161
nociception, 128, 129, 130, 175–176
Noddings, Nel, 44
Nuffield Council on Bioethics, 1, 2, 5, 14, 82, 86, 87, 88, 89, 90, 99, 100, 103, 105, 106, 107, 108, 109, 148
Nussbaum, Martha G., 57, 114, 115

pain, 2, 3, 7, 8, 11, 16–17, 20, 22, 23, 25–28, 30, 32, 33, 35, 36, 37, 38, 48, 53, 54, 63, 69, 70, 72, 73, 75–76, 77, 83, 103–104, 106–109, 111, 113, 114, 115, 116, 117, 121, 122, 123, 124, 128–131, 135, 136–137, 138–139, 140–141, 144, 145, 165, 166, 167, 168, 169, 173, 174, 175–176
Peters, Ted, 157
Petrinovich, Peter, 60, 66
playing God, 2, 157–158, 164
Post, Stephen G., 71
priority, 39, 42, 43, 44, 62, 67, 73, 111, 120–122, 125, 139, 143, 144–145, 156, 165, 166, 169
  strong human p., 13, 35, 39, 44, 45, 47, 49, 51, 52, 62, 74, 92, 104, 107, 125, 126, 139, 147
  weak human p., 11, 13, 15, 39, 44, 45, 47, 48, 49, 51, 52, 69, 70, 74, 76, 77, 78, 104, 106, 111, 124, 125, 126, 137, 138, 139, 144, 147, 150, 156, 179
pronuclear microinjection, 5, 7, 150, 151, 152, 153, 160, 166, 169, 174
prototype, 11, 12, 13, 14, 15, 17, 21, 28, 35, 39, 44, 45, 47, 49, 50, 51, 52, 53, 54, 59, 62, 63, 68, 69, 72, 73, 74, 78, 80, 82, 92, 104, 107, 111, 119, 124, 125, 126, 127, 128, 138, 139, 147, 148, 156

Radcliffe Richards, Janet, 66
Ramsey, Paul, 158
Rawls, John, 79, 82
reduction, 12, 39, 63, 75, 103, 104, 105, 106–108, 135, 139, 140, 144, 168, 169, 173
refinement, 12, 63, 75, 103, 104, 107, 108–109, 122, 173
Regan, Tom, 11, 13, 29–36, 39, 50, 51, 52, 53, 54, 56, 59, 63, 68, 69, 75, 76, 82, 92, 105, 108, 122, 125, 126, 127, 128, 132, 133, 164
relational ethics, 43, 44, 50, 51, 80
relational properties, 43, 44, 49, 54, 55, 63, 124
replacement, 6, 12, 26, 39, 63, 75, 87, 103, 104, 105–106, 107, 150, 173
respect, 15, 29, 30, 32, 33, 35, 38, 50, 80, 104, 139, 156
Rideout, William M., 170, 171
Röcklinsberg, Helena, 154, 155, 168, 169
Rollin, Bernard, 13, 14, 114, 117, 120, 122, 162–163
Rolston III, Holmes, 161, 164
Russell, W. M. S., 12, 63, 75, 103, 107, 108, 122
Rutgers, Bart, 69–70, 163
Ryder, Richard, 27, 106–107, 138–139, 141

Savage-Rumbaugh, Sue, 132
scientific value, 11, 75, 77, 85, 90–92, 95, 99, 100, 142
Segerdahl, Pär, 132
sentience, 12, 16, 25, 43, 54, 55, 63, 69, 74, 75, 127, 128, 130–133, 167, 173
Shanks, Niall, 2, 11, 14, 91–103
Sherlock, Richard, 2, 160
Sigmund, Curt D., 170

Singer, Peter, 11, 13, 21–29, 30, 32, 35, 36, 38, 39, 40, 41, 43, 44, 49, 50, 51, 53, 54, 56, 59, 61, 63, 68, 69, 75, 76, 82, 106, 108, 122, 125, 126, 127, 128, 131–132, 133
Sjöberg, Lennart, 9–10
Smith, John Maynard, 102
solidarity, 71–72, 80
special obligations, 11, 38, 43, 44, 49, 50, 54, 55, 59–60, 62, 63, 68, 71, 76, 133
species care, 11, 70, 72–74, 75, 76
speciesism, 27, 38, 40–42, 44, 56, 59, 67, 68, 73
Stafleu, F. R., 120, 139, 142, 143
Steel, Daniel, 93, 98, 99–100, 103
suffering, 1, 2, 3, 8, 11, 14, 17, 20, 21, 22, 25, 26, 28, 35, 38, 39, 44, 45, 47, 48, 49, 50, 51, 61, 68, 69, 72, 75, 77, 82, 102, 104, 106, 108, 109, 111–114, 115, 116, 123, 124, 125, 130–131, 136, 137, 138, 139, 140, 141, 142, 143, 144, 145, 147, 163, 166, 167, 168, 169

survey, 7, 8, 9, 10, 11, 58, 107, 155
Sweden, 3, 9, 47, 48, 76, 123, 154

Tannenbaum, Jerrold, 61, 62–63, 116, 117, 120
trade-off, 139, 140, 141, 143, 144, 145, 174, 175, 176, 177, 178, 179
transgenic animal, 7, 98, 151, 159

United Kingdom, 1, 3, 9, 10, 47, 48, 49, 123
United States, 3, 8, 47, 76
utilitarianism, 24, 25, 30, 33, 43, 51, 80, 123, 124, 138, 139

Verhey, Allen, 157, 158
Verhoog, Henk, 162

Warren, Mary Anne, 44
Welchman, Jennifer, 70–72

Zamir, Tzachi, 73, 76

VIBS

The **Value Inquiry Book Series** is co-sponsored by:

Adler School of Professional Psychology
American Indian Philosophy Association
American Maritain Association
American Society for Value Inquiry
Association for Process Philosophy of Education
Canadian Society for Philosophical Practice
Center for Bioethics, University of Turku
Center for Professional and Applied Ethics, University of North Carolina at Charlotte
Central European Pragmatist Forum
Centre for Applied Ethics, Hong Kong Baptist University
Centre for Cultural Research, Aarhus University
Centre for Professional Ethics, University of Central Lancashire
Centre for the Study of Philosophy and Religion, University College of Cape Breton
Centro de Estudos em Filosofia Americana, Brazil
College of Education and Allied Professions, Bowling Green State University
College of Liberal Arts, Rochester Institute of Technology
Concerned Philosophers for Peace
Conference of Philosophical Societies
Department of Moral and Social Philosophy, University of Helsinki
Gannon University
Gilson Society
Haitian Studies Association
Ikeda University
Institute of Philosophy of the High Council of Scientific Research, Spain
International Academy of Philosophy of the Principality of Liechtenstein
International Association of Bioethics
International Center for the Arts, Humanities, and Value Inquiry
International Society for Universal Dialogue
Natural Law Society
Philosophical Society of Finland
Philosophy Born of Struggle Association
Philosophy Seminar, University of Mainz
Pragmatism Archive at The Oklahoma State University
R.S. Hartman Institute for Formal and Applied Axiology
Research Institute, Lakeridge Health Corporation
Russian Philosophical Society
Society for Existential Analysis
Society for Iberian and Latin-American Thought
Society for the Philosophic Study of Genocide and the Holocaust
Unit for Research in Cognitive Neuroscience, Autonomous University of Barcelona
Whitehead Research Project
Yves R. Simon Institute

Titles Published

Volumes 1 - 179 see www.rodopi.nl

180.   Florencia Luna, *Bioethics and Vulnerability: A Latin American View*. A volume in **Values in Bioethics**

181.   John Kultgen and Mary Lenzi, Editors, *Problems for Democracy*. A volume in **Philosophy of Peace**

182.   David Boersema and Katy Gray Brown, Editors, *Spiritual and Political Dimensions of Nonviolence and Peace*. A volume in **Philosophy of Peace**

183.   Daniel P. Thero, *Understanding Moral Weakness*. A volume in **Studies in the History of Western Philosophy**

184.   Scott Gelfand and John R. Shook, Editors, *Ectogenesis: Artificial Womb Technology and the Future of Human Reproduction*. A volume in **Values in Bioethics**

185.   Piotr Jaroszyński, *Science in Culture*. A volume in **Gilson Studies**

186.   Matti Häyry, Tuija Takala, Peter Herissone-Kelly, Editors, *Ethics in Biomedical Research: International Perspectives*. A volume in **Values in Bioethics**

187.   Michael Krausz, *Interpretation and Transformation: Explorations in Art and the Self*. A volume in **Interpretation and Translation**

188.   Gail M. Presbey, Editor, *Philosophical Perspectives on the "War on Terrorism."* A volume in **Philosophy of Peace**

189.   María Luisa Femenías, Amy A. Oliver, Editors, *Feminist Philosophy in Latin America and Spain*. A volume in **Philosophy in Latin America**

190.   Oscar Vilarroya and Francesc Forn I Argimon, Editors, *Social Brain Matters: Stances on the Neurobiology of Social Cognition*. A volume in **Cognitive Science**

191.   Eugenio Garin, *History of Italian Philosophy*. Translated from Italian and Edited by Giorgio Pinton. A volume in **Values in Italian Philosophy**

192. Michael Taylor, Helmut Schreier, and Paulo Ghiraldelli, Jr., Editors, *Pragmatism, Education, and Children: International Philosophical Perspectives*. A volume in **Pragmatism and Values**

193. Brendan Sweetman, *The Vision of Gabriel Marcel: Epistemology, Human Person, the Transcendent*. A volume in **Philosophy and Religion**

194. Danielle Poe and Eddy Souffrant, Editors, *Parceling the Globe: Philosophical Explorations in Globalization, Global Behavior, and Peace*. A volume in **Philosophy of Peace**

195. Josef Šmajs, *Evolutionary Ontology: Reclaiming the Value of Nature by Transforming Culture*. A volume in **Central-European Value Studies**

196. Giuseppe Vicari, *Beyond Conceptual Dualism: Ontology of Consciousness, Mental Causation, and Holism in John R. Searle's Philosophy of Mind*. A volume in **Cognitive Science**

197. Avi Sagi, *Tradition vs. Traditionalism: Contemporary Perspectives in Jewish Thought*. Translated from Hebrew by Batya Stein. A volume in **Philosophy and Religion**

198. Randall E. Osborne and Paul Kriese, Editors, *Global Community: Global Security*. A volume in **Studies in Jurisprudence**

199. Craig Clifford, *Learned Ignorance in the Medicine Bow Mountains: A Reflection on Intellectual Prejudice*. A volume in **Lived Values: Valued Lives**

200. Mark Letteri, *Heidegger and the Question of Psychology: Zollikon and Beyond*. A volume in **Philosophy and Psychology**

201. Carmen R. Lugo-Lugo and Mary K. Bloodsworth-Lugo, Editors, *A New Kind of Containment: "The War on Terror," Race, and Sexuality*. A volume in **Philosophy of Peace**

202. Amihud Gilead, *Necessity and Truthful Fictions: Panenmentalist Observations*. A volume in **Philosophy and Psychology**

203. Fernand Vial, *The Unconscious in Philosophy, and French and European Literature: Nineteenth and Early Twentieth Century*. A volume in **Philosophy and Psychology**

204. Adam C. Scarfe, Editor, *The Adventure of Education: Process Philosophers on Learning, Teaching, and Research*. A volume in **Philosophy of Education**

205. King-Tak Ip, Editor, *Environmental Ethics: Intercultural Perspectives*. A volume in **Studies in Applied Ethics**

206. Evgenia Cherkasova, *Dostoevsky and Kant: Dialogues on Ethics*. A volume in **Social Philosophy**

207. Alexander Kremer and John Ryder, Editors, *Self and Society: Central European Pragmatist Forum*, Volume Four. A volume in **Central European Value Studies**

208. Terence O'Connell, *Dialogue on Grief and Consolation*. A volume in **Lived Values, Valued Lives**

209. Craig Hanson, *Thinking about Addiction: Hyperbolic Discounting and Responsible Agency*. A volume in **Social Philosophy**

210. Gary G. Gallopin, *Beyond Perestroika: Axiology and the New Russian Entrepreneurs*. A volume in **Hartman Institute Axiology Studies**

211. Tuija Takala, Peter Herissone-Kelly, and Søren Holm, Editors, *Cutting Through the Surface: Philosophical Approaches to Bioethics*. A volume in **Values in Bioethics**

212. Neena Schwartz: *A Lab of My Own*. A volume in **Lived Values, Valued Lives**

213. Krzysztof Piotr Skowroński, *Values and Powers: Re-reading the Philosophical Tradition of American Pragmatism*. A volume in **Central European Value Studies**

214. Matti Häyry, Tuija Takala, Peter Herissone-Kelly and Gardar Arnason, Editors, *Arguments and Analysis in Bioethics*. A volume in **Values in Bioethics**

215. Anders Nordgren, *For Our Children: The Ethics of Animal Experimentation in the Age of Genetic Engineering*. A volume in **Values in Bioethics**